Advances in Numerical Mathematics

Edited by
Prof. Dr. Dres. h. c. Hans Georg Bock
Prof. Dr. Dr. h. c. Wolfgang Hackbusch
Prof. Mitchell Luskin
Prof. Dr. Dr. h. c. Rolf Rannacher

T0238470

Andreas Potschka

A Direct Method for Parabolic PDE Constrained Optimization Problems

Springer Spektrum

Andreas Potschka
Heidelberg, Germany

Dissertation Heidelberg University, Germany, 2011

ISSN 1616-2994
ISBN 978-3-658-04475-6   ISBN 978-3-658-04476-3 (eBook)
DOI 10.1007/978-3-658-04476-3

The Deutsche Nationalbibliothek lists this publication in the Deutsche Nationalbibliografie;
detailed bibliographic data are available in the Internet at http://dnb.d-nb.de.

Library of Congress Control Number: 2013955622

Springer Spektrum
© Springer Fachmedien Wiesbaden 2014

Printed on acid-free paper

Springer Spektrum is a brand of Springer DE.
Springer DE is part of Springer Science+Business Media.
www.springer-spektrum.de

Acknowledgments

I submitted this mathematical doctoral thesis in August 2011 to the faculty of natural sciences and mathematics of Heidelberg University and successfully passed its disputation on December 19, 2011.

I acknowledge the support by Deutsche Forschungsgemeinschaft (DFG) within the Internationales Graduiertenkolleg 710 *Complex processes: Modeling, Simulation and Optimization*, under grant BO864/12-1 within the Schwerpunktprogramm 1253 *Optimization with Partial Differential Equations*, and within the *Heidelberg Graduate School for Mathematical and Computational Methods in the Sciences (HGS)*. I am also grateful for support by Bundesministerium für Bildung und Forschung (BMBF) under grant 03BONCHD. Furthermore I would like to acknowledge support by the European Commission within the project *Embedded Optimization for Resource Constrained Platforms (EMBOCON)*.

I would especially like to thank: my advisers Hans Georg Bock and Rolf Rannacher; my mentors Sebastian Sager and Johannes Schlöder; my collaborators Jan Van Impe, Sebastian Engell, Moritz Diehl, Stefan Körkel, Ekaterina Kostina, Achim Küpper, and Filip Logist; my research stay hosts Moritz Diehl, Andreas Griewank, Matthias Heinkenschloss, Nick Trefethen, and Fredi Tröltzsch; my colleagues Jan Albersmeyer, Dörte Beigel, Kathrin Hatz, Christian Hoffmann, Christian Kirches, Simon Lenz, Andreas Sommer, Andreas Schmidt, and Leo Wirsching; my rocking friends Tommy Binford, Eddie Castillo, Rick Castillo, Roger Fletcher, Sean Hardesty, Denis Ridzal; my very close friends Tom Kraus, Anita and Micha Strauß, Evelyn and Tobias Stuwe; my siblings Susanne and Manuel; and my parents Maria and Michael.

I am indebted to Falk Hante, Christian Kirches, Mario Mommer, Sebastian Sager, and Johannes Schlöder for valuable comments on a first draft of this thesis.

Andreas Potschka

Abstract

In this thesis we develop a numerical method based on Direct Multiple Shooting for Optimal Control Problems (OCPs) constrained by time-periodic Partial Differential Equations (PDEs). The proposed method features asymptotically optimal scale-up of the numerical effort with the number of spatial discretization points. It consists of a Linear Iterative Splitting Approach (LISA) within a Newton-type iteration with globalization on the basis of natural level functions. We investigate the LISA-Newton method in the framework of Bock's κ-theory and develop reliable a-posteriori κ-estimators. Moreover we extend the inexact Newton method to an inexact Sequential Quadratic Programming (SQP) method for inequality constrained problems and provide local convergence theory. In addition we develop a classical and a two-grid Newton-Picard preconditioner for LISA and prove grid-independent convergence of the classical variant for a model problem. Based on numerical results we can claim that the two-grid version is even more efficient than the classical version for typical application problems. Moreover we develop a two-grid approximation for the Lagrangian Hessian which fits well in the two-grid Newton-Picard framework and yields a reduction of 68 % in runtime for a nonlinear benchmark problem compared to the use of the exact Lagrangian Hessian. We show that the quality of the fine grid controls the accuracy of the solution while the quality of the coarse grid determines the asymptotic linear convergence rate, i.e., Bock's κ. Based on reliable κ-estimators we facilitate automatic coarse grid refinement to guarantee fast convergence. For the solution of the occurring large-scale Quadratic Programming Problems (QPs) we develop a structure exploiting two-stage approach. In the first stage we exploit the Multiple Shooting and Newton-Picard structure to reduce the large-scale QP to an equivalent QP whose size is independent of the number of spatial discretization points. For the second stage we develop extensions for a Parametric Active Set Method (PASM) to achieve a reliable and efficient solver for the resulting, possibly nonconvex QP. Furthermore we construct three illustrative, counter-intuitive toy examples which show that convergence of a one-shot one-step optimization method is neither necessary nor sufficient for the convergence of the forward problem method. For three regularization approaches to recover convergence our analysis shows that de-facto loss of convergence cannot be avoided with these approaches. We have further implemented the proposed methods within a code called MUSCOP which features automatic derivative generation for the model functions and dynamic system solutions of first and second order, parallelization on the Multiple Shooting structure, and a hybrid language programming paradigm to minimize setup and solution time for new application problems. We demonstrate the applicability, reliability, and efficiency of MUSCOP and thus the proposed numerical methods and techniques on a sequence of PDE OCPs of growing difficulty ranging from linear academic problems, over highly nonlinear academic problems of mathematical biology to a highly nonlinear real-world chemical engineering problem in preparative chromatography: The Simulated Moving Bed (SMB) process.

Contents

List of acronyms

AD	Algorithmic Differentiation
BDF	Backward Differentiation Formula
BVP	Boundary Value Problem
ECOP	Equality Constrained Optimization Problem
END	External Numerical Differentiation
FDM	Finite Difference Method
FEM	Finite Element Method
FVM	Finite Volume Method
IND	Internal Numerical Differentiation
IRAM	Implicitly Restarted Arnoldi Method
IVP	Initial Value Problem
KKT	Karush-Kuhn-Tucker
LICQ	Linear Independence Constraint Qualification
LISA	Linear Iterative Splitting Approach
MOL	Method Of Lines
NDGM	Nodal Discontinuous Galerkin Method
NLP	Nonlinear Programming Problem
NMT	Natural Monotonicity Test
OCP	Optimal Control Problem
ODE	Ordinary Differential Equation

OOP   Object Oriented Programming

PASM   Parametric Active Set Method

PCG   Preconditioned Conjugate Gradient

PDE   Partial Differential Equation

PQP   Parametric Quadratic Programming

QP   Quadratic Programming Problem

RMT   Restrictive Monotonicity Test

SCC   Strict Complementarity Condition

SMB   Simulated Moving Bed

SOSC   Second Order Sufficient Condition

SQP   Sequential Quadratic Programming

VDE   Variational Differential Equation

1 Introduction

"The miracle of the appropriateness of the language of mathematics for the formulation of the laws of physics is a wonderful gift which we neither understand nor deserve. We should be grateful for it and hope that it will remain valid in future research and that it will extend, for better or for worse, to our pleasure even though perhaps also to our bafflement, to wide branches of learning."

— E.P. WIGNER [164]

Mathematics today permeates an ever increasing part of the sciences far beyond mathematical physics just as about 50 years ago Nobel Prize laureate Wigner has hoped for. In particular mathematical methods for simulation and optimization of quantitative mathematical models continue to face growing demand in disciplines ranging from engineering, biology, economics, physics, etc. even to emerging areas of psychology or archeology (see, e.g., Sager et al. [139], Schäfer et al. [140]).

In this thesis we focus on mathematical and computational methods for the class of Optimal Control Problems (OCPs) pioneered by Pontryagin and Bellman in the middle of the 20th century. General mathematical optimization problems consist of finding a solution candidate which satisfies a set of constraints and minimizes a certain objective function. OCPs are optimization problems whose free variables comprise states and controls from (usually infinite dimensional) function spaces constrained to satisfy given differential equations. The differential equations describe the behavior of a dynamic system which can be controlled in a prescribed way.

The treatment of constraints given by Partial Differential Equations (PDEs) is one major challenge that we address in this thesis. PDEs appear when spatially distributed phenomena need to be taken into account, e.g., when we describe the diffusion of a substance in a liquid. Ordinary Differential Equations (ODEs), which describe the evolution of a system in time, are not a satisfactory mathematical tool for the description of spatial effects (although we shall use them to approximate solutions of PDEs). A considerable amount of theory and practical computational methods is available today for ODE OCPs. The presence of PDE constraints causes additional difficulties both on the theoretical as well as on the numerical side and is a much younger field of research especially in the aspect of methods which heavily rely on high computing power.

OCPs are inherently infinite problems because we seek solutions in function spaces. We can divide numerical methods for OCPs into two main classes: Direct and indirect methods. The defining line between the two is somewhat blurry, especially when we cross borders of mathematical communities. We base our classification here on the sequence of discretization and optimization. In indirect methods we first derive optimality conditions in function space which we discretize afterwards. In direct methods we discretize the problem first and then find an optimizer of the resulting Nonlinear Programming Problem (NLP). Moreover we often end up with an implicit characterization of the control via artificially introduced costate or adjoint variables in indirect methods. This is in contrast to direct methods for which the discretized control usually occurs explicitly as one or the only remaining variable. Indirect methods for ODE OCPs are mostly based on Dynamic Programming (see, e.g., Bellman [15]) or Pontryagin's Maximum Principle (see, e.g., Pontryagin et al. [126]). Tröltzsch [152] in his introductory textbook for PDE OCPs also treats only indirect methods. A discussion of direct and indirect methods for PDE OCPs is given in Hinze et al. [86, Chapter 3]. In the 1980's the endeavor to apply numerical optimization quickly to new application areas and new problems led to the development of direct methods for ODE OCPs, most notably collocation methods (see, e.g., Bär [8], Bock [23], Biegler [18]) and Direct Multiple Shooting (Bock [22], Plitt [125], Bock and Plitt [25]). One advantage of direct methods is that the optimality conditions of an NLP are generic, whereas optimality conditions of undiscretized OCPs need to be reestablished for each new problem and often require partial a-priori knowledge of the mathematical structure of the solution which in general is not available for many application problems. At the crux of creating an efficient direct optimization method is structure exploitation in the numerical solution of the NLP. Usually either Sequential Quadratic Programming (SQP) or Interior Point methods are employed (see, e.g., the textbook of Nocedal and Wright [121]). These iterative methods require the computation of derivatives of the objective function and the constraints. Derivative free methods (see, e.g., the introductory textbook by Conn et al. [34]) are typically not suited because of the high number of unknowns and because nonlinear constraints can only be treated with excessive computational effort.

It is our goal in this thesis to extend Direct Multiple Shooting for ODE OCPs in order to make it applicable and continue its success story for a class of PDE OCPs. The first hurdle on this venture is the large problem size of the discretized OCPs. Schäfer [141] describes in his dissertation approaches to address this difficulty by exploitation of the special mathematical structure of the discretized OCPs. His approaches lead to a reduction in the number of needed directional derivatives for the dynamical system. The Schäfer approach requires only a constant number of directional derivatives per optimization iteration while the number of directional derivatives for conventional Direct Multiple Shooting depends linearly on the number of spatial discretization points which typically grow prohibitively large.

However, the approach of Schäfer cannot be applied efficiently to OCPs with boundary conditions in time, the treatment of which is another declared goal of this thesis. PDE OCPs with time-periodicity conditions are even more difficult because for each spatial discretization point one additional constraint arises. In order to obtain an algorithm whose required number of directional derivatives is independent of the spatial discretization we have developed a globalized inexact SQP method in extension to ideas for inexact Newton methods (Ortega and Rheinboldt [122], Dembo et al. [41]), inexact SQP methods (Diehl et al. [48], Wirsching [166]), the Newton-Picard approach (Lust et al. [110]), and globalization via natural level functions (Bock [22], Bock et al. [26], Deuflhard [44]).

Boundary conditions in time occur often in practical applications, most of the time in form of a periodicity constraint. In this thesis we apply the investigated methods to the optimization of a real-world chromatographic separation process called Simulated Moving Bed (SMB). Preparative chromatography is one of various examples in the field of process operations for which periodic operation leads to a considerable increase in process performance compared to batch operation. The complicated structure of optimal solutions makes mathematical optimization an indispensable tool for the practitioner (see, e.g., Nilchan and Pantelides [120], van Noorden et al. [156], Toumi et al. [151], de la Torre et al. [40], Kawajiri and Biegler [93, 94], Agarwal et al. [1]).

1.1 Results of this thesis

For the first time we propose a method based on Direct Multiple Shooting for time-periodic PDE OCPs which features optimal scale-up of the effort in the number of spatial discretization points. This result is based on grid-independence of the number of inexact SQP iterations and a bound on the numerical effort for one inexact SQP iteration as essentially a constant times the effort for the solution of one Initial Value Problem (IVP) of the dynamical system. We can solve a nonlinear discretized large-scale optimization problem with roughly 700 million variables (counting intermediate steps of the IVP solutions as variables) in under half an hour on a current commodity desktop machine. Although developed particularly for PDE OCPs with time-periodicity constraints in mind, the method can also be applied to problems with fixed initial conditions instead of time-periodicity constraints and is thus considerably versatile.

Based on an inner Linear Iterative Splitting Approach (LISA) for the linear systems we review a LISA-Newton method. It is well-known that the linear asymptotic convergence rate of a LISA-Newton method with l inner LISA iterations coincides with the asymptotic convergence rate of the LISA method to the power of l. We prove this result for the first time in the framework of Bock's κ-theory. Truncated Neumann series occur in the proof which yield a closed form for the backward error of the inexact linear system solves. This backward error is of significant importance not only for the linear system itself but also in Bock's Local Contraction Theorem (Bock [24]) which characterizes the local convergence of Newton-type methods.

The previous result enables us to develop three novel a-posteriori κ-estimators which are computed from the iterates of the inner LISA iterations. We highlight the complications which result from the occurrence of non-diagonalizable iteration matrices from a geometrical point of view with examples.

We further extend LISA-Newton methods to SQP methods and prove that limit points satisfy a first order necessary optimality condition and that a second order sufficiency condition transfers from the Quadratic Programming Problem (QP) in the solution to the solution of the NLP. Moreover we describe the use of inexact Jacobians and Hessians within a generalized LISA method based on QPs. We also attempt an extension of a globalization strategy for LISA-Newton methods using natural level functions for the case of inexact SQP methods. We discuss important details of the numerical implementation and show that the developed strategy works reliably on numerical examples of practical relevance.

For LISA methods for time-periodic PDE OCPs we develop Newton-Picard preconditioners. We propose a classical variant based on Lust et al. [110] and a two-grid variant. We show that it is of paramount importance for numerical efficiency to modify the classical Newton-Picard preconditioner to use an L^2-based projector instead of a Euclidean projector. Moreover we prove grid-independent convergence of the classical Newton-Picard preconditioner on a linear-quadratic time-periodic PDE OCP. We further give numerical evidence that the two-grid variant is more efficient on a wide range of practical problems. For the extension of the proposed preconditioners to the nonlinear case for use in LISA-Newton methods we discuss several difficulties of the classical Newton-Picard preconditioner. We also develop a new two-grid Hessian approximation which fits naturally in the two-grid Newton-Picard framework and yields a reduction of 68 % in runtime for an exemplary nonlinear benchmark problem. Moreover we show that the two-grid Newton-Picard LISA-Newton method is scaling invariant. This property is of considerable importance for the reliability of the method on already badly conditioned problems.

The analysis reveals that the quality of the fine grid controls the accuracy of the solution while the quality of the coarse grid determines the asymptotic linear convergence rate, i.e., Bock's κ, of the two-grid Newton-Picard LISA-Newton method. Based on the newly established reliable a-posteriori κ-estimates we develop a numerical strategy for automatic determination of when to refine the coarse grid to guarantee fast convergence.

We further develop a structure exploiting two-stage strategy for the solution of QP subproblems in the inexact SQP method. The first stage is an extension of the traditional condensing step in SQP methods for Direct Multiple Shooting which exploits the constraint for periodicity or alternatively given fixed initial values for the PDE in addition to the Multiple Shooting matching conditions. This strategy reduces the large-scale QP to an equivalent QP whose size is independent of the spatial discretization. The reduction can be efficiently computed because it additionally exploits the (two-grid) Newton-Picard structure in the QP constraint and Hessian matrices. For the second stage we develop a Parametric Active Set Method (PASM) which can also treat nonconvex QPs with indefinite Hessian matrices. This capability is required because we want to treat nonconvex NLPs using accurate approximations for Lagrangian Hessians. We propose numerical techniques for improving the reliability of our PASM code which outperforms several other popular QP codes in terms of reliability.

The Newton-Picard LISA method can also be interpreted as a one-shot one-step approach for a linear PDE OCP. The almost optimal convergence theorem which we prove for the considered model problem supports the conjecture that such one-step approaches will in general yield optimization algorithms which converge as fast as the algorithm for the forward problem, which consists of satisfying the constraints for fixed controls. Contrary to common belief, however, we have constructed three small-scale, equality constrained QPs which illustrate that the convergence for the forward problem method is neither sufficient nor necessary for the convergence of the one-step optimization method. Furthermore we show that existing one-step techniques to enforce convergence might lead to de-facto loss of convergence with contraction factors of almost 1. These examples and results can serve as a warning signal or guiding principle for the choice of assertions which one might want to attempt to prove about one-step methods. It also justifies that we prove convergence of the Newton-Picard LISA only for a model problem.

We have put substantial effort into the implementation of the proposed ideas in a new software package called MUSCOP. Based on a hybrid programming design principle we strive to keep the code both easy to use and easy to maintain/develop further at the same time. The code features parallelization on the Multiple Shooting structure and automatic generation of derivatives of first and second order of the model functions and dynamic systems in order to reduce setup and solution time for new application problems to a minimum.

Finally we use MUSCOP to demonstrate the applicability, reliability, and efficiency of the proposed numerical methods and techniques on a sequence of PDE OCPs of growing difficulty: Linear and nonlinear boundary control tracking problems subject to the time-periodic linear heat equation in 2D and 1D, a tracking problem in bacterial chemotaxis which features a strong nonlinearity in the convective term, and finally a real-world practical example: Optimal control of the ModiCon variant of the SMB process.

1.2 Thesis overview

This thesis is structured in three parts: Theoretical foundations, numerical methods, and applications and numerical results. In Chapter 2 we give a short introduction to Bochner spaces and sketch the functional analytic setting for parabolic PDE in order to formulate the PDE OCP that serves as the point of origin for all further investigations in this thesis.

We present a direct optimization approach in Chapter 3. After a discussion of the discretize-then-optimize and optimize-then-discretize paradigms we describe a multi-stage discretization approach: Given a hierarchy of spatial discretizations we employ the Method Of Lines (MOL) to obtain a sequence of large-scale ODE OCPs which we subsequently discretize with Direct Multiple Shooting. We then formulate the resulting NLPs and discuss their numerical challenges.

In Chapter 4 we give a concise review of elements of finite dimensional optimization theory for completeness. This concludes Part 1, theoretical foundations.

We begin Part 2, numerical methods, with the development of a novel inexact SQP method in Chapter 5. We commence the discussion with Newton-type methods and present Bock's Local Contraction Theorem and its proof. Subsequently we review popular methods for globalization of Newton-type methods and discuss their limits when it comes to switching from globalized mode to fast local contraction mode. We then present the idea and several interpretations of globalization via natural level functions and explain how they overcome the problem of impediment of fast local convergence. The natural level function approach leads to computable monotonicity tests for the globalization strategy. A review of the Restrictive Monotonicity Test (RMT) and a Natural Monotonicity Test (NMT) for LISA-Newton methods then precedes an exhaustive discussion of the convergence of LISA and its connection with Bock's κ-theory. On this basis we develop three a-posteriori κ-estimators which are based on the LISA iterates. In addition we propose an extension to SQP methods, prove that a first-order necessary optimality condition holds if the method converges, and further show that a second order sufficiency condition transfers from the QP in the solution to the solution of the NLP. Finally we present a novel extension to inexact SQP methods on the basis of a generalized LISA for QPs.

In Chapter 6 we develop so-called Newton-Picard preconditioners for time-periodic OCPs. We discuss a classical and a two-grid projective approach. For the classical approach we show grid-independent convergence. We conclude the chapter with a discussion of the application of Newton-Picard preconditioning in a LISA-Newton method for nonlinear problems and for Multiple Shooting.

We present three counter-intuitive toy examples in Chapter 7 which show that convergence of the forward problem method is neither sufficient nor necessary for the convergence of a corresponding one-step one-shot optimization approach. We furthermore analyze regularization approaches which are designed to enforce one-step one-shot convergence and demonstrate that de-facto loss of convergence cannot be avoided via these techniques.

In Chapter 8 we discuss condensing of the occurring large-scale QPs to equivalent QPs whose size is independent of the number of spatial discretization points. We further develop efficient numerical exploitation of the Multiple Shooting and Newton-Picard structure. Moreover we propose a two-grid Hessian matrix approximation which fits well in the framework of the two-grid Newton-Picard preconditioners. As a final remark we show scaling invariance of the Newton-Picard LISA-Newton method for PDE OCPs.

The solution of the resulting medium-scale QPs via PASM is our subject in Chapter 9. We identify numerical challenges in PASMs and develop strategies to meet these challenges, in particular the techniques of drift correction and flipping bounds. Furthermore we implement these strategies in a code called rpasm and demonstrate that rpasm outperforms other popular QP solvers in terms of reliability on a well-known test set. We conclude the chapter with an extension to nonconvex QPs which can arise when employing the exact Lagrangian Hessian or the two-grid Newton-Picard Hessian approximation. The proposed PASM is also considerably efficient because it can be efficiently hot-started.

In Chapter 10 we review numerical methods for automatic generation of derivatives on the basis of Algorithmic Differentiation (AD) and Internal Numerical Differentiation (IND). Furthermore we address issues with a monitor strategy in implicit numerical integrators which can lead to violation of the IND principle for the example of a linear 1D heat equation. Then we conclude the chapter with a short account on the numerical effort of IND.

We dedicate Chapter 11 to the design of the software package MUSCOP. Issues we address include programming paradigms and description of the various software components and their complex orchestration necessary for smart structure exploitation. This concludes Part 2, numerical methods.

In Part 3 we present applications and numerical results which were generated with MUSCOP. Linear boundary control for the periodic 2D heat equation is in the focus of our presentation in Chapter 12. We give numerical evidence of the failure of Euclidean instead of L_2 projection in classical Newton-Picard preconditioners. In accordance with the proof of mesh-independent convergence we give several computational results for varying problem data and discuss why the two-grid variant is superior to the classical Newton-Picard preconditioner.

We extend the problem to nonlinear boundary control in 1D in Chapter 13 and discuss numerical self-convergence. We can show that employing the two-grid Hessian approximation leads to an overall reduction in computation time of 68 %. We discuss parallelization issues and compare runtimes for different discretizations of the control in time. In all cases we give detailed information about the runtime spent in different parts of the algorithm and show exemplarily that with above 95 % most of the runtime is required for system simulation and IND.

In Chapter 14 we present a tracking type OCP for a (non-periodic) bacterial chemotaxis model in 1D. The model is characterized by a highly nonlinear convective term. We demonstrate the applicability of the proposed methods also to this problem and discuss the self-convergence of the computation.

Chapter 15 is the last chapter of this thesis. In it we present the SMB process and explain a mathematical model for chromatographic columns. We then present numerical results for the ModiCon variant of the SMB process for real-world data. We obtain optimal solutions with an accuracy which has not been achieved before. This concludes Part 3 and this thesis.

Chapters 6, 7, 9, 12, and parts of Chapter 15 are based on own previously published work. For completeness we reprint partial excerpts here with adaptions to the unified nomenclature and structure of this thesis. We give the precise references to the respective articles at the beginning of each of these chapters.

Part I

Theoretical foundations

2 Problem formulation

The goal of this chapter is to introduce the Optimal Control Problem (OCP) formulation which serves as the point of origin for all further investigations in this thesis. To this end we recapitulate elements of the theory of parabolic Partial Differential Equations (PDEs) in Section 2.1 and present a system of PDEs coupled with Ordinary Differential Equations (ODEs) in Section 2.2. The coupled system is one of the constraints among additional boundary and path constraints for the OCP which we describe in Section 2.3. We emphasize the particular aspects in which our problem setting differs and extends the setting most often found in PDE constrained optimization.

2.1 Dynamical models described by Partial Differential Equations

We treat processes which are modeled by a state u distributed in space and evolving over time. The evolution of u is deterministic and described by PDEs. The behavior of the dynamical system can further be influenced by a time and possibly space dependent control q.

Nonlinear instationary PDEs usually do not have solutions in classical function spaces. We recapitulate the required definitions for Bochner spaces and vector-valued distributions necessary for formulations which have solutions in a weak sense. The presentation here is based on Dautray and Lions [37], Gajewski et al. [57], and Wloka [167]. We omit all proofs which can be found therein. Throughout this chapter let $\Omega \in \mathbb{R}^d$ be a bounded open domain with sufficiently regular boundary $\partial\Omega$, X be a Banach space, and $d\mu$ denote the Lebesgue measure in $\mathbb{R}^d$.

We assume that the reader is familiar with basic concepts of functional analysis (see, e.g., Dunford and Schwartz [50]). We denote with $L^p(\Omega), 1 \leq p \leq \infty$, the *Lebesgue space* of μ-measurable $\mathbb{R}$-valued functions whose absolute value to the p-th power has a bounded integral over Ω if $p < \infty$ or which are essentially bounded if $p = \infty$. Functions which coincide μ-almost everywhere are considered identical. With $W^{k,p}(\Omega)$, $k \geq 0, 1 \leq p < \infty$ we denote the *Sobolev space* of functions in $L^p(\Omega)$ whose distributional derivatives up to order k lie in $L^p(\Omega)$.

The spaces $L^p(\Omega), W^{k,p}(\Omega)$ endowed with their usual norms are Banach spaces. The spaces $H^k(\Omega) := W^{k,2}(\Omega)$ equipped with their usual scalar product are Hilbert spaces. The construction of $L^p(\Omega)$ and $W^{k,p}(\Omega)$ can be generalized to functions with values in Banach spaces:

Definition 2.1 (Bochner spaces). By $L^p(\Omega;X)$, $1 \leq p < \infty$, we denote the space of all measurable functions $v : \Omega \to X$ satisfying

$$\int_\Omega \|v\|_X^p \, d\mu < \infty.$$

We identify elements of $L^p(\Omega;X)$ which coincide μ-almost everywhere and equip $L^p(\Omega;X)$ with the norm

$$\|v\|_{L^p(\Omega;X)} = \left(\int_\Omega \|v\|_X^p \, d\mu \right)^{1/p}.$$

Now we proceed in the following way: Generally we are interested in weak solutions $u \in W$ in an appropriate Hilbert space $W \subset L^2((t_1,t_2) \times \Omega)$ with finite $t_1, t_2 \in \mathbb{R}$. Functions in $L^2((t_1,t_2) \times \Omega)$ need not even be continuous and hence we must exercise care to give well-defined meaning to derivatives and the traces $u(t_1,.)$ and $u(t_2,.)$. This is not trivial because altering u on any set of measure zero, e.g., $\{t_1,t_2\} \times \Omega$, yields the same u in $L^2((t_1,t_2) \times \Omega)$. The traces are important for the formulation of boundary value conditions. We address these issues concerning the state space in three steps. In a first step, we write

$$L^2((t_1,t_2) \times \Omega) = L^2((t_1,t_2); L^2(\Omega)),$$

i.e., we interpret u as an L^2 function in time with values in the space of L^2 functions in space. Second, we can formulate the time derivative du/dt of u via the concept of vectorial distributional derivatives.

Definition 2.2. Let Y be another Banach space. We denote the space of continuous linear mappings from X to Y with $\mathcal{L}(X,Y)$.

Definition 2.3. The space of vectorial distributions of the interval $(t_1,t_2) \subset \mathbb{R}$ with values in the Banach space X is denoted by

$$\mathscr{D}'((t_1,t_2);X) := \mathscr{L}(C^\infty([t_1,t_2];\mathbb{R}),X).$$

We can identify every $u \in L^2((t_1,t_2);X) \subset L^1((t_1,t_2);X)$ with a distribution $T \in \mathscr{D}'((t_1,t_2);X)$ via the Bochner integral

$$T\varphi = \int_{t_1}^{t_2} u(t)\varphi(t)dt \quad \text{for all } \varphi \in C^\infty([t_1,t_2];\mathbb{R}).$$

Definition 2.4. The k-th derivative of T is defined via

$$\frac{\mathrm{d}^k T}{\mathrm{d}t^k}\varphi = (-1)^k \int_{t_1}^{t_2} u(t)\varphi^{(k)}(t)\mathrm{d}t.$$

Thus, $\mathrm{d}T/\mathrm{d}t \in \mathscr{D}'((t_1,t_2);X)$. We assume now that $X \hookrightarrow Y$, where $\hookrightarrow$ denotes continuous embedding. Hence it holds that

$$\mathscr{D}'((t_1,t_2);X) \hookrightarrow \mathscr{D}'((t_1,t_2);Y),$$
$$L^p((t_1,t_2);X) \hookrightarrow L^p((t_1,t_2);Y).$$

Let $u \in L^2((t_1,t_2);X)$. We say that $\mathrm{d}u/\mathrm{d}t \in L^2((t_1,t_2);Y)$ if there exists $u' \in L^2((t_1,t_2),Y)$ such that

$$\int_{t_1}^{t_2} u'(t)\varphi(t)\mathrm{d}t = \frac{\mathrm{d}T}{\mathrm{d}t}\varphi = -\int_{t_1}^{t_2} u(t)\varphi^{(1)}(t)\mathrm{d}t \quad \text{for all } \varphi \in C^\infty([t_1,t_2];\mathbb{R}),$$

and we identify $\mathrm{d}u/\mathrm{d}t := u'$. We also use the abbreviation $\partial_t u := \mathrm{d}u/\mathrm{d}t$.

In the third step, let V and H be separable Hilbert spaces and let V^* denote the dual space of V. We assume throughout that (V,H,V^*) is a *Gelfand triple*

$$V \overset{\mathrm{d}}{\hookrightarrow} H \overset{\mathrm{d}}{\hookrightarrow} V^*,$$

i.e., the embeddings of V in H and $H = H^*$ in V^* are continuous and dense. Now we choose $X = V$ and $Y = V^*$ in order to define the space of L^2 functions over V with time derivatives in L^2 over the dual V^* according to

$$W(t_1,t_2) = \{u \in L^2((t_1,t_2);V) \mid \partial_t u \in L^2((t_1,t_2);V^*)\}.$$

Lemma 2.5. *The space $W(t_1,t_2)$ is a Hilbert space when endowed with the scalar product*

$$(u,v)_{W(t_1,t_2)} = \int_{t_1}^{t_2} (u(t),v(t))_V \, \mathrm{d}t + \int_{t_1}^{t_2} (\partial_t u(t), \partial_t v(t))_{V^*} \, \mathrm{d}t.$$

Proof. See Wloka [167, Satz 25.4]. □

Theorem 2.6. *We can alter every $u \in W(t_1,t_2)$ on a set of measure zero to obtain a function in $C^0([t_1,t_2];H)$. Furthermore, if we equip $C^0([t_1,t_2];H)$ with the norm of uniform convergence then*

$$W(t_1,t_2) \hookrightarrow C^0([t_1,t_2];H).$$

Proof. See Dautray and Lions [37, Chapter XVIII, Theorem 1]. □

Corollary 2.7. *For $u \in W(t_1, t_2)$ the traces $u(t_1), u(t_2)$ have a well-defined meaning in H (but not in V in general).*

For the control variables we assume $q \in L^2((t_1, t_2); Q)$ where $Q \subseteq L^2(\Omega)^{n_q}$ or $Q \subseteq L^2(\partial\Omega)^{n_q}$ for distributed or boundary control, respectively. We can then formulate the parabolic differential equation

$$\partial_t u(t) + A(q(t), u(t)) = 0, \tag{2.1}$$

with a nonlinear elliptic differential operator $A : Q \times V \to V^*$. In the numerical approaches which we present in Chapters 5 and 6 we exploit that A is an elliptic operator. We further assume that A is defined via a semilinear (i.e., linear in the last argument) form $a : (Q \times V) \times V \to \mathbb{R}$ according to

$$\langle A(q(t), u(t)), \varphi \rangle_{V^* \times V} = a(q(t), u(t), \varphi) \quad \text{for all } \varphi \in V. \tag{2.2}$$

We consider Initial Value Problems (IVPs), i.e., PDE (2.1) subject to $u(t_1) = u^0 \in H$. The question of existence, uniqueness, and continuous dependence of solutions on the problem data u^0 and q cannot be answered satisfactorily in a general setting. However, there are problem-dependent sufficient conditions (compare, e.g., Gajewski et al. [57] for the case $A(q(t), u(t)) = A_q(q(t)) + A_u(u(t))$). A thorough discussion of this question is beyond the focus of this thesis.

Example 1. For illustration we consider the linear heat equation with Robin boundary control and initial values

$$\partial_t u = \Delta u \quad \text{in } (0,1) \times \Omega, \tag{2.3a}$$

$$\partial_\nu u + \alpha u = \beta q \quad \text{on } (0,1) \times \partial\Omega, \tag{2.3b}$$

$$u\big|_{t=0} = u^0, \tag{2.3c}$$

where $\alpha, \beta \in L^\infty(\partial\Omega)$ and ∂_ν denotes the derivative in the direction of the outwards pointing normal ν on $\partial\Omega$. We choose $V = H^1(\Omega)$ and $H = L^2(\Omega)$. Multiplication with a test function $\varphi \in V$ and integration by parts transform equations (2.3a) and (2.3b) into

$$0 = \int_\Omega \partial_t u(t) \varphi - \int_\Omega (\Delta u(t)) \varphi \tag{2.4a}$$

$$= \int_\Omega \partial_t u(t) \varphi + \int_\Omega \nabla u(t)^{\mathsf{T}} \nabla \varphi - \int_{\partial\Omega} \left(\nabla u(t)^{\mathsf{T}} \nu \right) \varphi \tag{2.4b}$$

$$= \int_\Omega \partial_t u(t) \varphi + \int_\Omega \nabla u(t)^{\mathsf{T}} \nabla \varphi + \int_{\partial\Omega} \alpha u(t) \varphi - \int_{\partial\Omega} \beta q(t) \varphi \tag{2.4c}$$

$$=: \int_\Omega \partial_t u(t) \varphi + a(q(t), u(t), \varphi), \tag{2.4d}$$

which serves as the definition for the semilinear form a and the corresponding operator A. We immediately observe that a is even bilinear on $(Q \times V) \times V$ in this example.

2.2 Coupled ODEs and PDEs

In some applications, e.g., in chemical engineering, the models consist of PDEs which are coupled with ODEs. We denote the ODE states, which are not distributed in space, by $v \in C^0([t_1, t_2]; \mathbb{R}^{n_v})$. These states can for instance model the accumulation of mass of a chemical species at an outflow port of a chromatographic column (compare Chapter 15). We can formulate the coupled system of differential equations as

$$\partial_t u(t) = -A(q(t), u(t), v(t)), \qquad \dot{v}(t) = f^{\text{ODE}}(q(t), u(t), v(t)), \tag{2.5a}$$

where $f^{\text{ODE}} : Q \times H \times \mathbb{R}^{n_v}$, subject to initial or boundary value conditions in time. We restrict ourselves to an autonomous formulation because the non-autonomous case can always be formulated as system (2.5) by introduction of an extra ODE state $\dot{v}_i = 1$ with initial value $v_i(t_1) = t_1$.

The question of existence, uniqueness, and continuous dependence on the data for the solution of IVPs with the differential equations (2.5) is even more challenging than for PDE IVPs and must be investigated for restriced problem classes (e.g., when A is not dependent on the $v(t)$ argument). Again, a thorough discussion of this question exceeds the scope of this thesis.

2.3 The Optimal Control Problem

We now state the OCP which is the point of origin for all further investigations of this thesis:

$$\underset{\substack{q \in L^2((0,1);Q) \\ u \in W(0,1) \\ v \in C^0([0,1];\mathbb{R}^{n_v})}}{\text{minimize}} \quad \Phi(u(1), v(1)) \tag{2.6a}$$

$$\text{s.t.} \quad \partial_t u = -A(q(t), u(t), v(t)), \qquad t \in (0,1), \tag{2.6b}$$

$$\dot{v} = f^{\text{ODE}}(q(t), u(t), v(t)), \qquad t \in (0,1), \tag{2.6c}$$

$$(u(0), v(0)) = r^{\text{b}}(u(1), v(1)), \tag{2.6d}$$

$$r^{\text{c}}(q(t), v(t)) \geq 0, \qquad t \in (0,1), \tag{2.6e}$$

$$r^{\text{e}}(v(1)) \geq 0, \tag{2.6f}$$

with nonlinear functions

$$\Phi : H \times \mathbb{R}^{n_v} \to \mathbb{R}, \quad r^b : H \times \mathbb{R}^{n_v} \to H \times \mathbb{R}^{n_v},$$

$$r^c : Q \times \mathbb{R}^{n_v} \to \mathbb{R}^{n_r^c}, \quad r^e : \mathbb{R}^{n_v} \to \mathbb{R}^{n_r^e}.$$

We now discuss each line of OCP (2.6) in detail.

The objective function Φ in line (2.6a) is different from what is typically treated in PDE constrained optimization. Often, even for nonlinear optimal control problems, the objective functions are assumed to consist of a quadratic term for the states, e.g., L^2 tracking type in space or in the space-time cylinder, plus a quadratic Tychonoff-type regularization term for the controls (see, e.g., Tröltzsch [152]) of the type

$$\frac{1}{2} \int_0^1 \left\| u(t) - u^{\mathrm{desired}}(t) \right\|_H^2 \mathrm{d}t + \frac{\gamma}{2} \int_0^1 \|q(t)\|_Q^2 \,\mathrm{d}t. \tag{2.7}$$

We remark that tracking type problems with objective (2.7) on the space-time cylinder can always be cast in the form of OCP (2.6) by introduction of an additional ODE state variable v_i subject to

$$\dot{v}_i(t) = \left\| u(t) - u^{\mathrm{desired}}(t) \right\|_H^2 + \gamma \|q(t)\|_Q^2, \quad v_i(0) = 0,$$

with the choice $\Phi(u(1), v(1)) = v_i(1)/2$. The applications we are interested in, however, can have economical objective functions which are not of tracking type.

Constraints (2.6b) and (2.6c) determine the dynamics of the considered system. We have already described them in detail in Sections 2.1 and 2.2 of this chapter.

Initial or boundary value constraints are given by equation (2.6d). Typical examples are pure initial value conditions via constant

$$r^b(u(1), v(1)) := (u^0, v^0)$$

or periodicity conditions

$$r^b(u(1), v(1)) := (u(1), v(1)).$$

Compared to initial value conditions the presence of boundary value conditions makes it more difficult to use reduced approaches which rely on a solution operator for the differential equations mapping a control q to a feasible state u. Instead of solving one IVP, the solution operator would have to solve one Boundary Value Problem (BVP) which is in general both theoretically and numerically more difficult. Thus we avoid this *sequential* approach in favor of a *simultaneous* approach in which the intermediate control and state iterates of the method may be infeasible for equations (2.6b) through (2.6d). Of course feasibility must be attained in the optimal solution.

Inequality (2.6e) is supposed to hold for almost all $t \in (0,1)$ and can be used to formulate constraints on the controls and ODE states. We deliberately do not include PDE state constraints in the formulation which give rise to various theoretical difficulties and are currently a very active field of research. We allow for additional inequalities on the ODE states at the end via inequality (2.6f). In the context of chemical engineering applications, the constraints (2.6e) and (2.6f) can comprise flow rate, purity, throughput constraints, etc.

Problems with free time-independent parameters can be formulated within problem class (2.6) via introduction of additional ODE states v_i with vanishing time derivative $\dot{v}_i(t) = 0$. Although the software package MUSCOP (see Chapter 11) treats time-independent parameters explicitly, we refrain from elaborating on these issues in this thesis in order to avoid notational clutter.

OCP (2.6) also includes the cases of free start and end time via a time transformation, e.g., $\tau(t) = (1-t)\tau_1 + t\tau_2 \in [\tau_1, \tau_2], t \in [0,1]$. This case plays an important role in this thesis, e.g., in periodic applications with free period duration, see Chapter 15.

Concerning regularity of the functions involved in OCP (2.6), we take a pragmatic view point: We assume that the problem can be consistently discretized (along the lines of Chapter 3) and that the resulting finite dimensional optimization problem is sufficiently smooth on each discretization level to allow for employment of fast numerical methods (see Chapter 5).

3 Direct Optimization: Problem discretization

The goal of this chapter is to obtain a discretized version of OCP (2.6). We discuss a so-called direct approach and summarize its main advantages and disadvantages in Section 3.1 in comparison with alternative approaches. In Sections 3.2 and 3.3 we discretize OCP (2.6) in two steps. First we discretize in space and obtain a large-scale ODE constrained OCP which we then discretize in time to obtain a large-scale Nonlinear Programming Problem (NLP) presented in Section 3.5. The numerical solution of this NLP is the subject of Part II in this thesis.

3.1 Discretize-then-optimize approach

We follow a *direct approach* to approximate the infinite dimensional optimization problem (2.6) by finite dimensional optimality conditions by first discretizing the optimization problem to obtain an NLP (*discretize-then-optimize* approach). Popular alternatives are *indirect* approaches where infinite dimensional optimality conditions are formulated (*optimize-then-discretize* approaches). These two main routes are displayed in Figure 3.1.

To give a detailed list and comprehensive comparison of direct and indirect approaches for OCPs is beyond the scope of this thesis. A short comparison for ODE OCPs can be found, e.g., in Sager [138]. For PDE OCPs we refer the reader to Hinze et al. [86, Chapter 3]. Both direct and indirect approaches have various advantages and disadvantages which render one or the other more appropriate for a concrete problem instance at hand. For the sake of brevity we restrict ourselves to only state the main reason why we have decided to apply a direct approach: While the most important property of indirect methods is certainly that they provide the deepest insight into the mathematical structure of the solution of the particular problem, the insight usually comes at the cost of heavy mathematical analysis, which might be too time consuming or too difficult for a practitioner. The direct approach that we lay out in this chapter enjoys the advantage that it can be applied in a straight-forward, generic, and almost automatic way. We believe that this

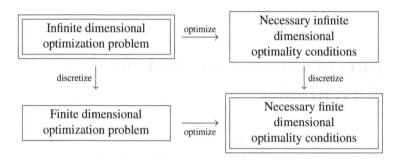

Figure 3.1: The two main routes to approximate an infinite dimensional optimization problem (upper left box) with necessary finite dimensional optimality conditions (lower right box) are direct approaches (discretize-then-optimize, lower left path) versus indirect approaches (optimize-then-discretize, upper right path).

property is paramount for successful deployment of the method in collaboration with practitioners.

We have chosen a Direct Multiple Shooting approach for the following numerical advantages: First, it has been shown in Albersmeyer [2] and Albersmeyer and Diehl [4] that Multiple Shooting can be interpreted as a *Lifted Newton method* which might reduce the nonlinearity of a problem and enlarge thus the domain of fast local convergence (see Chapter 5). Second, we can supply good initial value guesses—if available—for iterative solution methods through the local state variables. Third, we can employ advanced numerical methods for the solution and differentiation of the local IVPs (see Chapter 10). Fourth, due to the decoupling of the IVPs we can parallelize the solution of the resulting NLP on the multiple shooting structure (see Chapter 11).

3.2 Method Of Lines: Discretization in space

The first step in the discretization of OCP (2.6) consists of discretizing the function spaces V and Q, e.g., by Finite Difference Methods (FDMs), Finite Element Methods (FEMs), Finite Volume Methods (FVMs), Nodal Discontinuous Galerkin Methods (NDGMs), or spectral methods. Introductions to these methods are available in textbook form, e.g., LeVeque [109, 108], Braess [29], Hesthaven and Warburton [84], and Hesthaven et al. [85]. The approach of discretizing first in space and then in time is called Method Of Lines (MOL) and is often applied for parabolic problems (see, e.g., Thomée [150]). We must exercise care that the spatial discretization is appropriate for the concrete problem at hand. For instance, an FEM for advection dominated problems must be stabilized, e.g., by a Streamline Upwind Petrov Galerkin formulation (see, e.g., Brooks and Hughes [31]), and an NDGM for diffusion dominated problems must be stabilized by a jump penalty term (see, e.g., Warburton and Embree [161]).

We assume that after discretization we obtain a finite dimensional space $Q_h \subseteq Q$ and a hierarchy of finite dimensional spaces $V_h^l, l \in \mathbb{N}$, satisfying

$$V_h^1 \subset V_h^2 \subset \cdots \subset V.$$

We choose this setting for the following reasons: For many applications, especially in chemical engineering, an infinite dimensional control is virtually impossible to implement on a real process. In the case of room heating for instance, the temperature field is distributed in the three-dimensional room, but the degrees of freedom for the control will still be the scalar valued position of the radiator operation knob. In this case, a one-dimensional discretization Q_h of Q is fully sufficient. For the applications we consider, we always assume Q_h to be a low dimensional space. It is of paramount importance, however, to accurately resolve the system state u. For this reason, we assume that the spaces V_h^l are high dimensional for for large l. The numerical methods we describe in Part II will rely on and exploit these assumptions on the dimensionality of Q_h and V_h^l.

We can then use the finite dimensional spaces V_h^l and Q_h to obtain a discretization of the semilinear form a and thus the operator A from equation (2.2). On each level l we are led to an ODE of the form

$$M_u^l \dot{u}^l(t) = f^{\mathrm{PDE}(l)}(q(t), u^l(t), v(t)),$$

with symmetric positive-definite N_V^l-by-N_V^l matrix M_u^l, $u^l(t) \in \mathbb{R}^{N_V^l}, q(t) \in \mathbb{R}^{N_Q}$, and $f^{\mathrm{PDE}(l)} : \mathbb{R}^{N_Q} \times \mathbb{R}^{N_V^l} \times \mathbb{R}^{n_v} \to \mathbb{R}^{N_V^l}$. In this way we approximate PDE (2.1) with ODEs which are of large scale on finer discretization levels l. Let us illustrate this procedure in an example.

Example 2. We continue the example of the heat equation (Example 1) on the unit square $\Omega = (0,1)^2$ with boundary control. For the discretization in space we employ FEM. Let us assume that we have a hierarchy of nested triangular grids for the unit square (compare Figure 3.2) with vertices $\xi_i^l \in \Omega, i = 1, \ldots, N_V^l$, on level $l \in \mathbb{N}$. Let the set of triangular elements on level l be denoted by $\mathscr{T}^l$. We define the basis functions φ_i^l by requiring

$$\varphi_i^l(\xi_j^l) = \delta_{ij}, \quad \varphi_i^l \text{ is linear on each element } \mathscr{T} \in \mathscr{T}^l,$$

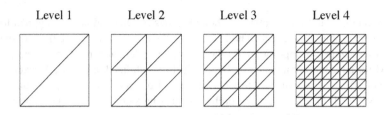

Figure 3.2: First four levels of an exemplary hierarchy of nested triangular meshes for the unit square obtained by uniform refinement with $N_V^l = (2^{l-1}+1)^2$ vertices on level $l = 1,\dots,4$.

with δ_{ij} denoting the Kronecker Delta. This construction yields the well-known hat functions. We then define the spaces V_h^l simply as the span of the basis functions φ_i^l, $i = 1,\dots,N_V^l$.

For the discretization of Q we assume that a partition of $\partial\Omega$ in segments $\mathscr{S}_k$, $k = 1,\dots,N_Q$, is given and choose Q_h as the span of their characteristic functions $\psi_k = \chi_{\mathscr{S}_k}$, which yields a piecewise constant discretization of the control on the boundary of the domain.

Let

$$u^l = \sum_{i=1}^{N_V^l} u_i^l \varphi_i^l \in V_h^l, \quad w^l = \sum_{i=1}^{N_V^l} w_i^l \varphi_i^l \in V_h^l, \quad q = \sum_{i=1}^{N_Q} q_i \psi_i \in Q_h$$

denote arbitrarily chosen discretized functions and their coordinates (in bold type-face) within their finite dimensional spaces. Then we obtain the following expressions for the terms occurring in equation (2.4) which allow for the evaluation of the integrals via matrices:

$$\int_\Omega u^l w^l = \sum_{i,j=1}^{N_V^l} u_i^l \left(\int_\Omega \varphi_i^l \varphi_j^l \right) w_j^l =: (u^l)^T M_V^l w^l,$$

$$\int_\Omega (\nabla u^l)^T \nabla w^l = \sum_{i,j=1}^{N_V^l} u_i^l \left(\int_\Omega (\nabla \varphi_i^l)^T \nabla \varphi_j^l \right) w_j^l =: (u^l)^T S^l w^l,$$

$$\int_{\partial\Omega} \alpha w^l q = \sum_{i=1}^{N_Q} \sum_{j=1}^{N_V^l} q_i \left(\int_{\partial\Omega} \alpha \psi_i \varphi_j^l \right) w_j^l =: q^T M_Q^l w^l,$$

$$\int_{\partial\Omega} \beta u^l w^l = \sum_{i,j=1}^{N_V^l} u_i^l \left(\int_{\partial\Omega} \beta \varphi_i^l \varphi_j^l \right) w_j^l =: (u^l)^T M_\partial^l w^l.$$

The occurring matrices are all sparse because each basis function has by construction a support of only a few neighboring elements. We now substitute $u(t), \varphi$ and $q(t)$ in equation (2.4) with their discretized counterparts to obtain

$$\dot{u}^l(t)^{\mathrm{T}} M_V^l e_i^l = -u^l(t)^{\mathrm{T}} S^l e_i^l - u^l(t)^{\mathrm{T}} M_\partial^l e_i^l + q(t)^{\mathrm{T}} M_Q^l e_i^l, \quad \text{for } i = 1, \dots, N_V^l,$$

where e_i^l denotes the i-th column of the N^l-by-N^l identity matrix. Exploiting symmetry of M_V^l, S^l, and M_∂^l yields the equivalent linear ODE formulation

$$M_V^l \dot{u}^l(t) = (-S^l - M_\partial^l) u^l(t) + (M_Q^l)^{\mathrm{T}} q(t) =: f^{\mathrm{PDE}(l)}(q(t), u^l(t), v(t)). \quad (3.1)$$

It is well-known that the state mass matrix on the left hand side of equation (3.1) is symmetric positive definite. To multiply equation (3.1) from the left with the dense matrix $(M_V^l)^{-1}$ is often avoided in order to preserve sparsity of the right hand side matrices.

We want to conclude Example 2 with the remark that in a FVM or an NDGM, where the basis functions are discontinuous over element boundaries, the mass matrix has block diagonal form and hence sparsity is preserved for $(M_V^l)^{-1}$. For spectral methods, all occurring matrices are usually dense anyway. In both cases, the inverse mass matrix is usually formulated explicitly in the right hand side of equation (3.1).

3.3 Direct Multiple Shooting: Discretization in time

After approximation of PDE (2.1) with large-scale ODEs as we have described in Section 3.2, we can employ Direct Multiple Shooting (see the seminal paper of Bock and Plitt [25]) to discretize the ODE constrained OCP. The aim of this section is to give an overview of Direct Multiple Shooting.

To this end let

$$0 = t^0 < \cdots < t^{n_{\mathrm{MS}}} = 1$$

denote a partition of the time interval $[0, 1]$, the so-called *shooting grid*. We further employ a piecewise discretization of the semi-discretized control $q(t)$ such that $q(t)$ is constant on the *shooting intervals*

$$I^i = (t^{i-1}, t^i), \quad i = 1, \dots, n_{\mathrm{MS}},$$

with values

$$q(t) = \sum_{i=1}^{n_{\mathrm{MS}}} q^{i-1} \chi_{I^i}(t).$$

Piecewise higher order discretizations in time are also possible, as long as the shooting intervals stay decoupled. Otherwise we loose the possibility for structure exploitation which is important for numerical efficiency reasons as we discuss in Chapter 8. In this thesis we restrict ourselves to piecewise constant control discretizations in time for reasons of simplicity.

We now introduce artificial initial values $(s^{l,i}, v^i), i = 0, \ldots, n_{MS}$, for the semi-discretized PDE states $u^l(t)$ and the ODE states $v(t)$, respectively. We define

$$f^{ODE(l)}(q(t), u^l(t), v(t)) := f^{ODE}(\sum_{j=1}^{N_Q} q_j(t) \psi_j, \sum_{j=1}^{N_V^l} u_j^l(t) \varphi_j, v(t))$$

and assume that each local IVP

$$M_u^l \dot{u}^l(t) = f^{PDE(l)}(q^{i-1}, u^l(t), v(t)), \qquad t \in I^i, \qquad u^l(t^{i-1}) = s^{l,i-1}, \qquad (3.2a)$$

$$\dot{v}(t) = f^{ODE(l)}(q^{i-1}, u^l(t), v(t)), \qquad t \in I^i, \qquad v(t^{i-1}) = v^{i-1}. \qquad (3.2b)$$

has a unique solution, denoted by the pair

$$(\overline{u}^{l,i}(t; q^{i-1}, s^{l,i-1}, v^{i-1}), \overline{v}^{l,i}(t; q^{i-1}, s^{l,i-1}, v^{i-1})).$$

Local existence and uniqueness of $(\overline{u}^{l,i}, \overline{v}^{l,i})$ are guaranteed by the Picard-Lindelöf theorem if the functions $f^{PDE(l)}$ and $f^{ODE(l)}$ are Lipschitz continuous in the second and third argument. By means of $(\overline{u}^{l,i}, \overline{v}^{l,i})$ we obtain a piecewise, finite dimensional parametrization of the state trajectories. To recover continuity of the entire trajectory across the shooting grid nodes we have to impose *matching conditions*

$$\begin{pmatrix} \overline{u}^{l,i}(t^i; q^{i-1}, s^{l,i-1}, v^{i-1}) \\ \overline{v}^{l,i}(t^i; q^{i-1}, s^{l,i-1}, v^{i-1}) \end{pmatrix} - \begin{pmatrix} s^{l,i} \\ v^i \end{pmatrix} = 0, \quad i = 1, \ldots, n_{MS}.$$

Remark 3.1. We introduce an additional artificial control variable $q^{n_{MS}}$ on the last shooting grid node in order to have the same structure of degrees of freedom in each t_i, $i = 0, \ldots, n_{MS}$. We shall always require $q^{n_{MS}} = q^{n_{MS}-1}$. This convention simplifies the presentation and implementation of the structure exploitation that we present in Chapter 8.

Remark 3.2. It is also possible and numerically advantageous to allow for different spatial meshes on each shooting interval I^i in combination with a-posteriori mesh refinement, see Hesse [83]. In that case the matching conditions have to be formulated differently. This topic, however, is beyond the scope of this thesis. We restrict ourselves to uniformly refined meshes which are equal for all shooting intervals.

3.4 Discretization of path constraints

Before we can formulate the discretized optimization problem we have been aiming at in this chapter, we need to repeat on each level l the construction of $f^{\mathrm{ODE}(l)}$ from f^{ODE} for the remaining functions

$$\Phi^l(s^{l,n_{\mathrm{MS}}},v^{n_{\mathrm{MS}}}) := \Phi(\textstyle\sum_{j=1}^{N_V^l} s_j^{l,n_{\mathrm{MS}}}\varphi_j, v^{n_{\mathrm{MS}}}),$$

$$r^{\mathrm{b}(l)}(s^{l,n_{\mathrm{MS}}},v^{l,n_{\mathrm{MS}}}) := r^{\mathrm{b}}(\textstyle\sum_{j=1}^{N_V^l} s_j^{l,n_{\mathrm{MS}}}\varphi_j, v^{l,n_{\mathrm{MS}}}),$$

$$r^{\mathrm{i}}(q(t),v(t)) := r^{\mathrm{c}}(\textstyle\sum_{j=1}^{N_Q} q_j(t)\psi_j, v(t)).$$

We observe that the path constraint containing r^{i} is supposed to hold in infinitely many points $t \in [0,1]$. There are different possibilities to discretize such a constraint (see Potschka [127] and Potschka et al. [128]). For the applications we treat in this thesis it is sufficient to discretize path constraint (2.6e) on the shooting grid

$$r^{\mathrm{i}}(q^{i-1},v^{i-1}) \geq 0, \quad i=1,\ldots,n_{\mathrm{MS}}.$$

3.5 The resulting Nonlinear Programming Problem

Finally we arrive at a finite dimensional optimization problem on each spatial discretization level l

$$\underset{(q^i,s^{l,i},v^i)_{i=0}^{n_{\mathrm{MS}}}}{\text{minimize}} \quad \Phi^l(s^{l,n_{\mathrm{MS}}},v^{n_{\mathrm{MS}}}) \tag{3.3a}$$

$$\text{s.t.} \quad r^{\mathrm{b}(l)}(s^{l,n_{\mathrm{MS}}},v^{l,n_{\mathrm{MS}}}) - (s^{l,0},v^{l,0}) = 0, \tag{3.3b}$$

$$\bar{u}^{l,i}(t^i;q^{i-1},s^{l,i-1},v^{i-1}) - s^{l,i} = 0, \quad i=1,\ldots,n_{\mathrm{MS}}, \tag{3.3c}$$

$$\bar{v}^{l,i}(t^i;q^{i-1},s^{l,i-1},v^{i-1}) - v^i = 0, \quad i=1,\ldots,n_{\mathrm{MS}}, \tag{3.3d}$$

$$q^{n_{\mathrm{MS}}} - q^{n_{\mathrm{MS}}-1} = 0, \tag{3.3e}$$

$$r^{\mathrm{i}}(q^{i-1},v^{i-1}) \geq 0, \quad i=1,\ldots,n_{\mathrm{MS}}, \tag{3.3f}$$

$$r^{\mathrm{e}}(v^{n_{\mathrm{MS}}}) \geq 0. \tag{3.3g}$$

Throughout we assume that all discretized functions are sufficiently smooth to apply the numerical methods of Part II. This includes that all functions need to be at least twice continuously differentiable. In the case of the functions $f^{\mathrm{PDE}(l)}$ and $f^{\mathrm{ODE}(l)}$ we might even need higher regularity to allow for efficient adaptive error control in the numerical integrator.

For the efficient solution of NLP (3.3) we have developed an inexact Sequential Quadratic Programming (SQP) method which we describe in Part II. We conclude this chapter with a summary of the numerical challenges:

Large scale. The NLPs have $n_{\mathrm{NLP}(l)} = (n_{\mathrm{MS}} + 1)\left(N_V^l + n_v\right) + n_{\mathrm{MS}}N_Q$ variables and are thus considered large-scale for finer levels l. The numerical methods which we describe in Part II aim at the efficient treatment of NLP (3.3) for large $N_V^l \approx 10^5$. The number of shooting intervals n_{MS} will be between 10^1 and 10^2 which amounts to an overall problem size $n_{\mathrm{NLP}(l)} \approx 10^7$. We want to remark that this calculation does not include the values of $\bar{u}^{l,i}$ which have to be computed in intermediate time steps between shooting nodes. There can be between 10^1 and 10^2 time steps per interval.

Efficient derivative generation. It is inevitable for the solution of large-scale optimization problems to use derivative-based methods. Hence we need numerical methods which deliver consistent derivatives of the functions occurring in NLP (3.3), especially in the matching conditions (3.3c) and (3.3d). In Chapter 10 we describe such a method which efficiently computes consistent derivatives of first and second order in an automated way.

Structure exploitation. Efficient numerical methods must exploit the multiple shooting structure of NLP (3.3). We present a *condensing* approach in Chapter 8 which operates on linearizations of NLP (3.3). Furthermore we develop preconditioners in Chapter 6 which exploit special spectral properties of the shooting Wronksi matrices. These spectral properties arise due to ellipticity of the operator A.

Mesh-independent local convergence. One of the main results of this thesis is that these preconditioners lead to mesh-independent convergence of the inexact SQP method, i.e., the number of iterations is asymptotically bounded by a reasonably small constant for $l \to \infty$. We prove this assertion for a model problem in Chapter 6 and the numerical results that we have obtained on the application problems in Part III suggest that this claim also holds for difficult real-world problems.

Global convergence. Often there is only few a-priori information available about the solution of real-world problems. Hence it is paramount to enforce convergence of the inexact SQP method also from starting points far away from the solution. However, it must be ensured that an early transition to fast local convergence is preserved. We describe such a globalization strategy based on *natural level functions* in Chapter 5.

4 Elements of optimization theory

In this short chapter we consider the NLP

$$\underset{x\in\mathbb{R}^n}{\text{minimize}} \quad f(x) \tag{4.1a}$$

$$\text{s.t.} \quad g_i(x) = 0, \quad i \in \mathscr{E}, \tag{4.1b}$$

$$g_i(x) \geq 0, \quad i \in \mathscr{I}, \tag{4.1c}$$

where $f : \mathbb{R}^n \to \mathbb{R}$ and $g : \mathbb{R}^n \to \mathbb{R}^m$ are twice continuously differentiable functions and the sets $\mathscr{E}$ and $\mathscr{I}$ form a partition of $\{1,\ldots,m\} =: \overline{m} = \mathscr{E} \,\dot\cup\, \mathscr{I}$. In the case of $\mathscr{E} = \overline{m}$, $\mathscr{I} = \{\}$, NLP (4.1) is called Equality Constrained Optimization Problem (ECOP).

4.1 Basic definitions

We follow Nocedal and Wright [121] in the presentation of the following basic definitions.

Definition 4.1. The set

$$\mathscr{F} = \{x \in \mathbb{R}^n \mid g_i(x) = 0, i \in \mathscr{E}, g_i(x) \geq 0, i \in \mathscr{I}\}$$

is called *feasible set*.

Definition 4.2. A point $x \in \mathscr{F}$ is called *feasible point*.

Definition 4.3. A point $x^* \in \mathbb{R}^n$ is called *global solution* if $x^* \in \mathscr{F}$ and

$$f(x^*) \leq f(x) \quad \text{for all } x \in \mathscr{F}.$$

Definition 4.4. A point $x^* \in \mathbb{R}^n$ is called *local solution* if $x^* \in \mathscr{F}$ and if there exists a neighborhood $U \subset \mathbb{R}^n$ of x^* such that

$$f(x^*) \leq f(x) \quad \text{for all } x \in U \cap \mathscr{F}.$$

Most algorithms for NLP (4.1) do not guarantee to return a global solution, which is virtually impossible if $\mathscr{F}$ is of high dimensionality as is the case in PDE constrained optimization problems. Thus we restrict ourselves to finding only local solutions. Research papers on global optimization can be found in Floudas and Pardalos [56] and in the *Journal of Global Optimization*.

4.2 Necessary optimality conditions

The numerical algorithms to be described in Part II for approximation of a local solution of NLP (4.1) are based on finding an approximate solution to necessary optimality conditions. We present these conditions after the following required definitions.

Definition 4.5. The *active set* at a feasible point $x \in \mathscr{F}$ is defined as

$$\mathscr{A}(x) = \{i \in \overline{m} \mid g_i(x) = 0\}.$$

Definition 4.6. The Linear Independence Constraint Qualification (LICQ) holds at $x \in \mathscr{F}$ if the the active constraint gradients $\nabla g_i(x)$, $i \in \mathscr{A}(x)$, are linearly independent.

Remark 4.7. There are also weaker constraint qualifications (see, e.g., Nocedal and Wright [121]). For our purposes it is convenient to use the LICQ.

Definition 4.8. The *Lagrangian* function is defined by

$$\mathscr{L}(z) = f(x) - \sum_{i \in \overline{m}} y_i g_i(x),$$

where $z := (x, y) \in \mathbb{R}^{n+m}$.

The following necessary optimality conditions are also called Karush-Kuhn-Tucker (KKT) conditions [92, 99].

Theorem 4.9 (First-Order Necessary Optimality Conditions). *Suppose that $x^* \in \mathbb{R}^n$ is a local solution of NLP (4.1) and that the LICQ holds at x^*. Then there is a Lagrange multiplier vector $y^* \in \mathbb{R}^m$ such that the following conditions are satisfied at $z^* = (x^*, y^*)$:*

$$\nabla_x \mathscr{L}(x^*, y^*) = 0, \tag{4.2a}$$

$$g_i(x^*) = 0, \quad i \in \mathscr{E}, \tag{4.2b}$$

$$g_i(x^*) \geq 0, \quad i \in \mathscr{I}, \tag{4.2c}$$

$$y_i^* \geq 0, \quad i \in \mathscr{I}, \tag{4.2d}$$

$$y_i^* g_i(x^*) = 0, \quad i \in \overline{m}. \tag{4.2e}$$

Proof. See Nocedal and Wright [121]. □

Remark 4.10. The Lagrange multipliers y are also called *dual* variables, in contrast to the *primal* variables x. We call $z^* = (x^*, y^*)$ primal-dual solution.

In the next definition we characterize a property which is favorable for the determination of the active set in a numerical algorithm because small changes in the problem data will not lead to changes in the active set at the solution.

Definition 4.11. Suppose $z^* = (x^*, y^*)$ is a local solution of NLP (4.1) satisfying (4.2). We say that the *Strict Complementarity Condition (SCC)* holds if $y_i > 0$ for all $i \in \mathscr{I} \cap \mathscr{A}(x^*)$.

A useful sufficient optimality condition is based on the notion of two cones:

Definition 4.12. Let $x \in \mathscr{F}$. The cone of linearized feasible directions is defined by

$$\mathscr{F}_1(x) = \{d \in \mathbb{R}^n \mid d^T \nabla g_i(x) = 0, i \in \mathscr{E}, \quad d^T \nabla g_i(x) \geq 0, i \in \mathscr{A}(x) \cap \mathscr{I}\}.$$

Definition 4.13. Let $x \in \mathscr{F}$. The cone of critical directions is defined by

$$\mathscr{C}(x, y) = \{d \in \mathscr{F}_1(x) \mid d^T \nabla g_i(x) = 0, \text{ for all } i \in \mathscr{A}(x) \cap \mathscr{I} \text{ with } y_i > 0\}.$$

The cone of critical directions plays an important role in the following sufficient optimality condition.

Theorem 4.14. *Let (x^*, y^*) satisfy the KKT conditions (4.2). If furthermore the Second Order Sufficient Condition (SOSC)*

$$d^T \nabla_{xx}^2 \mathscr{L}(x^*, y^*) d > 0 \quad \text{for all } d \in \mathscr{C}(x^*, y^*) \setminus \{0\}$$

holds then x^* is a strict local solution.

Proof. See Nocedal and Wright [121]. □

Part II

Numerical methods

5 Inexact Sequential Quadratic Programming

In this chapter we develop a novel approach for the solution of inequality constrained optimization problems. We first describe inexact Newton methods in Section 5.1 and investigate their local convergence in Section 5.2. In Section 5.3 we review strategies for the globalization of convergence and explain a different approach based on generalized level functions and monotonicity tests. An example in Section 5.4 illustrates the shortcomings of globalization strategies which are not based on the so called *natural level function*. We review the Restrictive Monotonicity Test (RMT) in Section 5.5 and propose a Natural Monotonicity Test (NMT) for Newton-type methods based on a Linear Iterative Splitting Approach (LISA). This combined approach allows for estimation of the critical constants which characterize convergence. We finally present how these results can be extended to global inexact SQP methods. We present efficient numerical solution techniques of the resulting sequence of Quadratic Programming Problems (QPs) in Chapters 8 and 9.

5.1 Newton-type methods

We consider the problem of finding a zero of a nonlinear function $F : D \subseteq \mathbb{R}^N \to \mathbb{R}^N$ which we assume to be continuously differentiable with Jacobian denoted by J. This case is important for computing local solutions of an ECOP because its KKT conditions (4.2) reduce to a system of nonlinear equations

$$F(z) := \begin{pmatrix} \nabla \mathcal{L}(z) \\ g(x) \end{pmatrix} = 0,$$

where $N = n + m$ and $z = (x, y)$ are the compound primal-dual variables. We shall discuss extensions for the inequality constrained case in Section 5.7.

The numerical methods of choice for the solution of $F(z) = 0$ are Newton-type methods: Given an initial solution guess z^0, we iterate

$$\Delta z^k = -M(z^k)F(z^k), \quad z^{k+1} = z^k + \alpha_k \Delta z^k, \tag{5.1}$$

with scalars $\alpha_k \in (0,1]$ and matrices $M(z) \in \mathbb{R}^{N \times N}$. The scalar α_k is a *damping* or *underrelaxation* parameter. We shall see in Sections 5.2 and 5.3 that the choice of $\alpha_k = 1$ is necessary for fast convergence close to a solution but choices $\alpha_k < 1$ are necessary to achieve convergence from initial guesses z^0 which are not sufficiently close to a solution.

Different choices of M lead to different Newton-type methods. Typical and important choices for M include *Quasi-Newton methods* (based on secant updates, see, e.g., Nocedal and Wright [121]), the *Simplified Newton method* ($M(z) = J^{-1}(z^0)$), and the *Newton method* ($M(z) = J^{-1}(z)$), provided the inverses exist. We have developed a method which uses Linear Iterative Splitting Approach (see Section 5.6.2) with a Newton-Picard deflation preconditioner (described in Chapter 6) to evaluate M.

Before we completely dive into the subject we want to clarify the naming of methods. We use *Newton-type method* as a collective term to refer to methods which can be cast in the form of equation (5.1). If the linearized subproblems are solved by an iterative procedure we use the term *inexact Newton method*. Unfortunately the literature is not consistent here: The often cited paper by Dembo et al. [41] uses *inexact Newton method* in the sense of our *Newton-type method* and *Newton-iterative method* in the sense of our *inexact Newton method* to distinguish between solving

$$\tilde{J}(z^k)\Delta z^k = -F(z^k) \quad \text{or} \quad J(z^k)\Delta z^k = -F(z^k) + r_k, \tag{5.2}$$

where $r_k \in \mathbb{R}^N$ is a residual and $\tilde{J}(z^k) \approx J(z^k)$. Some authors, e.g., Morini [115], further categorize *inexact Newton-like methods* which solve

$$\tilde{J}(z^k)\Delta z^k = -F(z^k) + r_k$$

in each step. From the point of view that equations (5.2) are merely characterizing the backward and the forward error of Δz^k for $J(z^k)\Delta z^k = -F(z^k)$, we believe that equations (5.2) should not be the basis for categorizing algorithms but rather be kept in mind for the analysis of all Newton-type methods. The following lemma shows that one can move from one interpretation to the other:

Lemma 5.1. *Let Δz^* solve $J(z^k)\Delta z^* = -F(z^k)$. If Δz^k is given via*

$$\Delta z^k = -M(z^k)F(z^k) \quad or \quad \tilde{J}(z^k)\Delta z^k = -F(z^k)$$

then the residual can be computed as

$$r_k = J(z^k)\Delta z^k + F(z^k).$$

Conversely, if r_k and Δz^k are given and $\left\| \Delta z^k \right\|_2 > 0$ then one possible $\tilde{J}(z^k)$ is given by

$$\tilde{J}(z^k) = J(z^k) - \frac{r_k (\Delta z^k)^{\mathrm{T}}}{(\Delta z^k)^{\mathrm{T}} \Delta z^k}.$$

Moreover, if $J(z^k)$ is invertible and $(\Delta z^k)^{\mathrm{T}} \Delta z^* \neq 0$ then

$$M(z^k) = \tilde{J}(z^k)^{-1} = \left(\mathbb{I} + \frac{(\Delta z^k - \Delta z^*)(\Delta z^k)^{\mathrm{T}}}{(\Delta z^k)^{\mathrm{T}} \Delta z^*} \right) J(z^k)^{-1}.$$

Proof. The first assertion is immediate. The second assertion can be shown via

$$\tilde{J}(z^k)\Delta z^k = -F(z^k) + r_k - r_k \frac{(\Delta z^k)^{\mathrm{T}} \Delta z^k}{(\Delta z^k)^{\mathrm{T}} \Delta z^k} = -F(z^k).$$

By virtue of the Sherman-Morrison-Woodbury formula (see, e.g., Nocedal and Wright [121]) we obtain

$$M(z^k) = \tilde{J}(z^k)^{-1} = J(z^k)^{-1} + \frac{J(z^k)^{-1} \frac{r_k (\Delta z^k)^{\mathrm{T}}}{(\Delta z^k)^{\mathrm{T}} \Delta z^k} J(z^k)^{-1}}{1 - \frac{(\Delta z^k)^{\mathrm{T}}}{(\Delta z^k)^{\mathrm{T}} \Delta z^k} J(z^k)^{-1} r_k}.$$

The last assertion then follows from $J(z^k)^{-1} r_k = \Delta z^k - \Delta z^*$. $\square$

In this thesis we focus on computing Δz^k iteratively via the iteration

$$\Delta z_{i+1}^k = \Delta z_i^k - \hat{M}(z^k)(J(z^k)\Delta z_i^k + F(z^k)), \tag{5.3}$$

with $\hat{M}(z^k) \in \mathbb{R}^{N \times N}$. We call iteration (5.3) Linear Iterative Splitting Approach (LISA) to emphasize that the iteration (which we further discuss in Section 5.6.2) is linear and based on a splitting

$$J(z^k) = \hat{J}(z^k) - \Delta J(z^k), \quad \hat{M}(z^k) = \hat{J}(z^k)^{-1}.$$

In this thesis $\hat{J}(z^k)$ will be given by a Newton-Picard preconditioner (see Chapter 6). For other possible choices of $\hat{J}(z^k)$ in this context, which include Jacobi, Gauß-Seidel, Successive Overrelaxation, etc., we refer the reader to Ortega and Rheinboldt [122] and Saad [135]. There is no consistent naming convention available in the literature: We can find names like *generalized linear iterations* (Ortega and Rheinboldt [122]) or *basic linear methods* (Saad [135]) for what we call LISA. A characterization of $M(z^k)$ based on $\hat{M}(z^k)$ for LISA in terms of a truncated Neumann series shall be given in Lemma 5.27.

A linear iteration like (5.3) can in principle be accelerated by the use of Krylov-space methods at the cost of making the iteration nonlinear. We abstain from nonlinear acceleration in this thesis because the Newton-Picard preconditioners are already powerful enough when used without acceleration (see Chapter 6).

In the following sections we review the theory for local and global convergence of Newton-type methods.

5.2 Local convergence

We present a variant of the Local Contraction Theorem (see Bock [24]). Let the set of Newton pairs be defined according to

$$\mathcal{N} = \{(z, z') \in D \times D \,|\, z' = z - M(z)F(z)\}$$

and let $\|.\|$ denote a norm of $\mathbb{R}^N$. We need two conditions on J and M:

Definition 5.2 (Lipschitz condition: ω-condition). The Jacobian J together with the approximation M satisfy the ω-condition in D if there exists $\omega < \infty$ such that for all $t \in [0,1], (z, z') \in \mathcal{N}$

$$\left\| M(z') \left(J(z + t(z' - z)) - J(z) \right) (z - z') \right\| \le \omega t \left\| z - z' \right\|^2.$$

Definition 5.3 (Compatibility condition: κ-condition). The approximation M satisfies the κ-condition in D if there exists $\kappa < 1$ such that for all $(z, z') \in \mathcal{N}$

$$\left\| M(z')(\mathbb{I} - J(z)M(z))F(z) \right\| \le \kappa \left\| z - z' \right\|.$$

Remark 5.4. If $M(z)$ is invertible, then the κ-condition can also be written as

$$\left\| M(z')(M^{-1}(z) - J(z))(z - z') \right\| \le \kappa \left\| z - z' \right\|, \quad \forall (z, z') \in \mathcal{N}.$$

With the constants from the previous two definitions we define

$$c_k = \kappa + (\omega/2) \left\| \Delta z^k \right\|$$

and for $c_0 < 1$ the closed ball

$$D_0 = \overline{B}(z^0; \left\| \Delta z^0 \right\| / (1 - c_0)).$$

The following theorem characterizes the local convergence of a full step (i.e., $\alpha_k = 1$) Newton-type method in a neighborhood of the solution. Because of its importance we include the well-known proof.

Theorem 5.5 (Local Contraction Theorem). *Let J and M satisfy the ω-κ-conditions in D and let $z^0 \in D$. If $c_0 < 1$ and $D_0 \subseteq D$, then $z^k \in D_0$ and the sequence (z^k) converges to some $z^* \in D_0$ with convergence rate*

$$\left\| \Delta z^{k+1} \right\| \le c_k \left\| \Delta z^k \right\| = \kappa \left\| \Delta z^k \right\| + (\omega/2) \left\| \Delta z^k \right\|^2.$$

Furthermore, the a-priori estimate

$$\left\| z^{k+j} - z^* \right\| \le \frac{(c_k)^j}{1 - c_k} \left\| \Delta z^k \right\| \le \frac{(c_0)^{k+j}}{1 - c_0} \left\| \Delta z^0 \right\|$$

holds and the limit z^* satisfies

$$M(z^*)F(z^*) = 0.$$

If additionally $M(z)$ is continuous and nonsingular in z^*, then

$$F(z^*) = 0.$$

Proof based on the Banach Fixed Point Theorem. The assumption $c_0 < 1$ and the Definition of D_0 imply that $z^0, z^1 \in D_0$. We assume that $z^{k+1} \in D_0$. Then

$$
\begin{aligned}
\left\| \Delta z^{k+1} \right\| &= \left\| M(z^{k+1})F(z^{k+1}) \right\| \\
&= \| M(z^{k+1}) \left(F(z^k) - J(z^k)M(z^k)F(z^k) \right) \\
&\quad + M(z^{k+1}) \left(F(z^{k+1}) - F(z^k) + J(z^k)M(z^k)F(z^k) \right) \| \\
&\le \kappa \left\| \Delta z^k \right\| + \left\| M(z^{k+1}) \left(\int_0^1 \frac{dF}{dt}(z^k + t\Delta z^k)dt - J(z^k)\Delta z^k \right) \right\| \\
&\le \kappa \left\| \Delta z^k \right\| + \int_0^1 \left\| M(z^{k+1}) \left(J(z^k + t\Delta z^k) - J(z^k) \right) \Delta z^k \right\| dt \\
&\le \kappa \left\| \Delta z^k \right\| + (\omega/2) \left\| \Delta z^k \right\|^2 = c_k \left\| \Delta z^k \right\|.
\end{aligned}
$$

It follows that the sequence (c_k) is monotonically decreasing because

$$c_{k+1} = \kappa + \frac{\omega}{2} \left\| \Delta z^{k+1} \right\| \le \kappa + c_k \frac{\omega}{2} \left\| \Delta z^k \right\| = c_k - (1 - c_k)\frac{\omega}{2} \left\| \Delta z^k \right\| \le c_k.$$

Telescopic application of the triangle inequality yields $z^{k+2} \in D_0$ due to

$$\left\| z^{k+2} - z^0 \right\| \le \sum_{j=0}^{k+1} \left\| \Delta z^j \right\| \le \sum_{j=0}^{k+1} (c_0)^j \left\| \Delta z^0 \right\| \le \frac{\left\| \Delta z^0 \right\|}{1 - c_0}.$$

By induction we obtain $z^k \in D_0$ for all $k \in \mathbb{N}$. From

$$\left\| z^{k+j} - z^k \right\| = \sum_{i=k}^{k+j-1} \left\| \Delta z^i \right\| \leq \sum_{i=0}^{j-1} (c_0)^k \left\| \Delta z^i \right\| \leq (c_0)^k \frac{\left\| \Delta z^0 \right\|}{1 - c_0}$$

follows that (z^k) is a Cauchy sequence and thus converges to a fixed point $z^* \in D_0$. For the a-priori estimate consider

$$\left\| z^{k+j} - z^* \right\| \leq \sum_{i=0}^{\infty} \left\| \Delta z^{k+j+i} \right\| \leq \sum_{i=0}^{\infty} (c_k)^i \left\| \Delta z^{k+j} \right\| \leq \frac{(c_k)^j}{1 - c_k} \left\| \Delta z^k \right\|.$$

In the limit

$$z^* = z^* - M(z^*)F(z^*) \quad \Rightarrow \quad M(z^*)F(z^*) = 0$$

holds which shows the remaining assertions. $\square$

Remark 5.6. If F is linear we obtain $\omega = 0$ and if furthermore $M(z)$ is constant the convergence theory is completely described by Theorem 5.26 to be presented.

Remark 5.7. Assume J is nonsingular throughout D_0. Then the full step Newton method with $M(z) = J^{-1}(z)$ converges quadratically in D_0 due to $\kappa = 0$.

Remark 5.8. In accordance with Deuflhard's *algorithmic paradigm* (see Deuflhard [44]) we assume the constants κ and ω to be the infimum of all possible candidates which satisfy the inequalities in their respective definitions. These values are in general computationally unavailable. Within the algorithms to be presented we approximate the infima from below with computational estimates denoted by $[\kappa]$ and $[\omega]$ by sampling the inequalities over a finite subset $\widetilde{\mathscr{N}} \subset \mathscr{N}$ which comprises various iterates of the algorithm.

5.3 Globalization of convergence

Most strategies for enforcing global convergence of inexact SQP methods are based on globalization techniques like trust region (see, e.g., Heinkenschloss and Vicente [82], Heinkenschloss and Ridzal [81], Walther [158], Gould and Toint [65]) or line search (see, e.g., Biros and Ghattas [20], Byrd et al. [33]). The explicit algorithmic control of Jacobian approximations is usually enforced via an adaptively chosen termination criterion for an inner preconditioned Krylov solver for the solution of the linearized system. In some applications, efficient preconditioners are available which cluster the eigenvalues of the preconditioned system and thus effectively reduce the number of inner Krylov iterations necessary to solve the linear system exactly (see Battermann and Heinkenschloss [9], Battermann and Sachs [10], Biros and Ghattas [19]).

We shall show in Section 5.4 that both line search and trust region methods can lead to unnecessarily damped iterates in the vicinity of the solution where fast local convergence in the sense of the Local Contraction Theorem 5.5 is already possible.

Our aim in this section is to present the theoretical tools to understand this undesirable effect and to introduce the idea of monotonicity tests.

We begin the discussion on the basis of the Newton method and extend it to Newton-type methods in Section 5.6 and to inexact SQP methods for inequality constrained optimization problems in Section 5.7.

5.3.1 Generalized level functions

It is well known that the local Newton method with $\alpha_k = 1$ is affine invariant under linear transformations in the residual and variable space:

Lemma 5.9. *Let $A, B \in \mathrm{GL}(N)$. Then the iterates z^k for the Newton method on $F(z)$ and the iterates $\tilde{z}^k$ for*

$$\tilde{F}(\tilde{z}) := AF(B\tilde{z})$$

with $\tilde{z}^0 := B^{-1} z^0$ are connected via

$$\tilde{z}^k = B^{-1} z^k, \quad k \in \mathbb{N}.$$

Proof. Let $k \in \mathbb{N}$. We assume $\tilde{z}^k = B^{-1} z^k$, obtain

$$\tilde{z}^{k+1} = \tilde{z}^k - \tilde{J}(\tilde{z}^k)^{-1} \tilde{F}(\tilde{z}^k) = \tilde{z}^k - B^{-1} J(B\tilde{z}^k)^{-1} A^{-1} AF(B\tilde{z}^k) = B^{-1} z^{k+1},$$

and complete the proof by induction. □

It is desirable to conserve at least part of the invariance for the determination of the damping parameters α_k. Our goal here are globalization strategies which are invariant under linear transformations in the residual space with $A \in \mathrm{GL}(N), B = \mathbb{I}$. This type of invariance is called *affine covariance* (see Deuflhard [44] for a classification of invariants for local and global Newton-type methods). We shall elaborate in Section 5.4 why affine covariance is important for problems which exhibit high condition numbers of the Jacobian $J(z^*)$ in the solution. This is the typical case in PDE constrained optimization problems.

We can immediately see that the Lipschitz constant ω in Definition 5.2 and the compatibility constant κ in Definition 5.3 are indeed independent of A. Thus they lend themselves to be used in an affine invariant globalization strategy.

We conclude this section with a descent result for the Newton direction on *generalized level functions*

$$T(z|A) := \tfrac{1}{2} \|AF(z)\|_2^2, \quad A \in \mathrm{GL}(N).$$

Generalized level functions extend the concept of the *classical level function $T(z|\mathbb{I})$* and play an important role in affine invariant globalization strategies. The following simple but nonetheless remarkable lemma (see, e.g., Deuflhard [44]) shows that the Newton direction is a direction of descent for all generalized level functions.

Lemma 5.10. *Let $F(z) \neq 0$. Then, for all $A \in \mathrm{GL}(N)$, the Newton increment $\Delta z = -J(z)^{-1}F(z)$ satisfies*

$$\Delta z^{\mathrm{T}} \nabla T(z|A) = -2T(z|A) < 0.$$

Proof. $\Delta z^{\mathrm{T}} \nabla T(z|A) = -F(z)^{\mathrm{T}} J(z)^{-\mathrm{T}} \left(F(z)^{\mathrm{T}} A^{\mathrm{T}} A J(z) \right)^{\mathrm{T}} = -2T(z|A) < 0.$ □

However, decrease $T(z + \alpha \Delta z|A) < T(z|A)$ might only be valid for $\alpha \ll 1$. We shall illustrate this problem with an example in Section 5.4. For the construction of efficient globalization strategies, A must be chosen such that the decrease condition is valid for a maximal range of α, as we shall discuss in Sections 5.5 and 5.6.

5.3.2 The Newton path

The Newton path plays a fundamental role in affine invariant globalization strategies for the Newton method. We present two characterizations of the Newton path, one geometrical and one based on a differential equation.

For preparation let $A \in \mathrm{GL}(N)$ and define the level set associated with $T(z|A)$ according to

$$G(z|A) := \{z' \in D \subseteq \mathbb{R}^N \mid T(z'|A) \leq T(z|A)\}.$$

Iterative monotonicity (descent) with respect to $T(z|A)$ can the be written in the form

$$z^{k+1} \in \mathring{G}(z^k|A), \quad \text{if } \mathring{G}(z^k|A) \neq \{\}.$$

To achieve affine invariance of the globalization strategy, descent must be independent of A. Thus we define

$$\overline{G}(z) := \bigcap_{A \in \mathrm{GL}(N)} G(z|A).$$

The geometric derivation of the Newton path due to Deuflhard [42, 43, 44] and the connection to the continuous analog of the Newton method characterized by Davidenko [38] is given by the following result.

Theorem 5.11. *Let $J(z)$ be nonsingular for all $z \in D$. For some $\widehat{A} \in \mathrm{GL}(N)$ let the path-connected component of $G(z^0|\widehat{A})$ in z^0 be compact and contained in D. Then the path-connected component of $\overline{G}(z^0)$ is a topological path $\overline{z} : [0,2] \to \mathbb{R}^N$, the so-called Newton path, which satisfies*

$$F(\overline{z}(\alpha)) = (1-\alpha)F(z^0), \tag{5.4a}$$

$$T(\overline{z}(\alpha)|A) = (1-\alpha)^2 T(z^0|A) \quad \forall A \in \mathrm{GL}(N), \tag{5.4b}$$

$$\frac{\mathrm{d}\overline{z}}{\mathrm{d}\alpha}(\alpha) = -J(\overline{z}(\alpha))^{-1}F(z^0), \tag{5.4c}$$

$$\overline{z}(0) = z^0, \quad \overline{z}(1) = z^*, \tag{5.4d}$$

$$\left.\frac{\mathrm{d}\overline{z}}{\mathrm{d}\alpha}\right|_{\alpha=0} = -J(\overline{z}^0)^{-1}F(z^0) = \Delta z^0. \tag{5.4e}$$

Proof. See Deuflhard [44, Theorem 3.1.4]. □

Remark 5.12. The differential equation (5.4c) is derived from the homotopy

$$H(z,\alpha) = F(z) - (1-\alpha)F(z^0) = 0, \tag{5.5}$$

which gives rise to the function $\overline{z}(\alpha)$ upon invocation of the Implicit Function Theorem. After solving equation (5.5) for $F(z^0)$ and using the reparametrization $\alpha(t) = 1 - \exp(-t)$ we can recover the so-called *continuous Newton method* or *Davidenko differential equation* (Davidenko [38])

$$J(\overline{z}(t))\dot{\overline{z}}(t) = -F(\overline{z}(t)), \quad t \in [0,\infty), \overline{z}(0) = z^0. \tag{5.6}$$

Theorem 5.11 justifies that the Newton increment Δz^k is a distinguished direction not only locally but also far away from a solution. It might only be too large, hence necessitating the need for damping through α_k.

In other words, the Newton path is the set of points generated from infinitesimal Newton increments (denoted by $\dot{\overline{z}}(t)$ instead of Δz^k). Performing nonzero steps $\alpha_k \Delta z^k$ in iteration (5.1) for the Newton method gives rise to a different Newton path emanating from each $z^k, k \in \mathbb{N}$. It seems inevitable that for a proof of global convergence based on the Newton path the iterates z^k must be related to a single Newton path, which we discuss in Section 5.5.

5.3.3 The natural level function and the Natural Monotonicity Test

In this section we assemble results for the natural level function and the *NMT*. For the presentation we follow Section 3.3 of Deuflhard [44]. At first, we restrict the presentation to the Newton method. Extensions to Newton-type methods with iterative methods for the linear algebra shall be discussed in Section 5.6. The main purpose of this section is the presentation of the natural level functions, defined by

$$T_k^*(z) := T(z|J(z^k)^{-1}),$$

which have several attractive properties. We can already observe that descent in T_k^* can be evaluated by testing for natural monotonicity

$$\left\|\overline{\Delta z}^{k+1}\right\| < \left\|\Delta z^k\right\|,$$

where one potential step of the Simplified Newton method can be used to evaluate $\overline{\Delta z}^{k+1}$ according to

$$J(z^k)\overline{\Delta z}^{k+1} = -F(z^{k+1}).$$

As already mentioned, the Lipschitz constant ω plays a fundamental role in the global convergence theory based on generalized level functions. In contrast to Bock [24], however, Deuflhard [44] uses a different definition for the Lipschitz constant:

Definition 5.13 ($\widehat{\omega}$-condition for the Newton method). The Jacobian J satisfies the $\widehat{\omega}$-condition in D if there exists $\widehat{\omega} < \infty$ such that for all $(z, z') \in D \times D$

$$\left\|J(z)^{-1}\left(J(z') - J(z)\right)(z' - z)\right\| \leq \widehat{\omega}\left\|z' - z\right\|^2.$$

Remark 5.14. In order to compare the magnitudes of ω and $\widehat{\omega}$ we define the set of interpolated Newton pairs

$$\mathcal{N}_t = \{(z, \widetilde{z}) \in D \times D \mid t \in [0,1], (z, z') \in \mathcal{N}, \widetilde{z} = z + t(z' - z), z \neq \widetilde{z}\}.$$

Then we can compute the smallest ω according to

$$
\begin{aligned}
\omega &= \sup_{\substack{(z,z') \in \mathcal{N}, z \neq z' \\ t \in (0,1]}} \frac{\left\|J(z')^{-1}\left(J(z + t(z' - z)) - J(z)\right)t(z' - z)\right\|}{t^2\left\|z' - z\right\|^2} \\
&= \sup_{(z,\widetilde{z}) \in \mathcal{N}_t} \frac{\left\|J(z - J(z)^{-1}F(z))^{-1}\left(J(\widetilde{z}) - J(z)\right)(\widetilde{z} - z)\right\|}{\left\|\widetilde{z} - z\right\|^2},
\end{aligned}
$$

which coincides with Definition 5.13 of $\widehat{\omega}$ except for the evaluation of the weighting matrix in the Lipschitz condition at a different point. Because $\mathcal{N}_t$ is much smaller than $D \times D$, the constant $\widehat{\omega}$ must be expected to be much larger than ω.

This will lead to smaller bounds on the step sizes α_k. Furthermore, in practical computations with a Newton-type method, only ω can be estimated efficiently from the iterates because $\widehat{\omega}$ is not explicitly restricted to the set of Newton pairs $\mathscr{N}$.

Remark 5.15. Most proofs which rely on $\widehat{\omega}$ can also be carried out in a similar fashion with ω but some theoretical results cannot be stated as elegantly. As an example, the occurrence of the condition number $\text{cond}(AJ(z^k))$ in Theorem 5.16 relies on using $\widehat{\omega}$. We take up the pragmatic position that ω should be used for all practical computations but we also recede to $\widehat{\omega}$ if we can gain theoretical insight about qualitative convergence behavior of Newton-type methods.

The first theorem which relies on $\widehat{\omega}$ characterizes step sizes α_k which yield optimal values for a local descent estimate of generalized level functions $T(z|A)$ in the Newton method.

Theorem 5.16. *Let D be convex, $J(z)$ nonsingular for all $z \in D$. Let furthermore J satisfy the $\widehat{\omega}$-condition in D, $z^k \in D, A \in \text{GL}(N)$, and $G(z^k|A) \subset D$. Let Δz^k denote the Newton direction and define the* Kantorovich quantities

$$h_k := \left\| \Delta z^k \right\| \widehat{\omega}, \quad \overline{h}_k := h_k \text{cond}(AJ(z^k)).$$

Then we obtain for $\alpha \in [0, \min(1, 2/\overline{h}_k(A))]$ that

$$\left\| AF(z^k + \alpha \Delta z^k) \right\| \leq t_k(\alpha|A) \left\| AF(z^k) \right\|,$$

where

$$t_k(\alpha|A) := 1 - \alpha + (1/2)\alpha^2 \overline{h}_k(A).$$

The optimal choice of the damping factor in terms of this local estimate is

$$\overline{\alpha}_k(A) := \min(1, 1/\overline{h}_k(A)).$$

Proof. See Deuflhard [44, Theorem 3.12]. □

Theorem 5.16 lends itself to the following global convergence theorem.

Theorem 5.17. *In addition to the assumptions of Theorem 5.16 let D_0 denote the path-connected component of $G(z^0|A)$ in z^0 and assume that $D_0 \subseteq D$ is compact. Then the damped Newton iteration with damping factors*

$$\alpha_k \in [\varepsilon, 2\overline{\alpha}_k(A) - \varepsilon]$$

and sufficiently small D_0-dependent $\varepsilon > 0$ converges to a solution point z^0.

Proof. See Deuflhard [44, Theorem 3.13]. □

Theorem 5.17 is a theoretical result for global convergence based on descent in any Generalized level function $T(z|A)$ with fixed A. However, the "optimal" step size chosen according to Theorem 5.16 is reciprocally proportional to the condition number $\mathrm{cond}(AJ(z^k))$. Thus a choice of A far away from $J(z^k)^{-1}$, e.g., $A = \mathbb{I}$ on badly conditioned problems, will lead to quasi-stalling of the globalized Newton method even within the domain of local contraction. Such a globalization strategy is practically useless for difficult problems, even though there exists a proof of global convergence.

This observation has led to the development of natural level functions $T_k^* = T(z|J(z^k)^{-1})$. The choice of $A_k = J(z^k)^{-1}$ yields the optimal value of

$$1 \le \mathrm{cond}_2(A_k J(z^k)) = 1,$$

and thus the largest value for the step size $\overline{\alpha}_k$. As already mentioned at the beginning of this section, we recall that descent in T_k^* can be evaluated by the NMT

$$\left\| \overline{\Delta z}^{k+1} \right\| < \left\| \Delta z^k \right\|,$$

where $\overline{\Delta z}^{k+1}$ is the increment for one potential Simplified Newton step.

Natural level functions have several outstanding properties as stated by Deuflhard [44, Section 3.3.2]:

Extremal properties For $A \in \mathrm{GL}(N)$ the reduction factors $t_k(\alpha|A)$ and the theoretical optimal damping factors $\overline{\alpha}_k(A)$ satisfy

$$t_k(\alpha|A_k) = 1 - \alpha + (1/2)\alpha^2 h_k \le t_k(\alpha|A),$$
$$\overline{\alpha}_k(A_k) = \min(1, 1/h_k) \ge \overline{\alpha}_k(A).$$

Steepest descent property The steepest descent direction for $T(z|A)$ in z^k is

$$-\nabla T(z^k|A) = -(AJ(z^k))^T A F(z^k).$$

With $A = A_k$ we obtain

$$\Delta z^k = -\nabla T(z^k|A_k),$$

which means that the damped Newton method in z^k is a method of steepest descent for the natural level function T_k^*.

Merging property Full steps and thus fast local convergence are guaranteed in proximity of the solution

$$\left\|\Delta z^k\right\|_2 \leq 1/\widehat{\omega} \quad \Rightarrow \quad h_k \leq 1 \quad \Rightarrow \quad \overline{\alpha}_k(A_k) = 1.$$

Asymptotic distance function For $F \in C^2(D)$, we verify that

$$T(z|J(z^*)^{-1}) = \tfrac{1}{2}\|z - z^*\|_2^2 + \mathcal{O}(\|z - z^*\|_2^3).$$

Hence the NMT asymptotically estimates monotonicity in the distance to the solution. The use of A_k can be considered a nonlinear preconditioner.

However, a straight-forward proof of global convergence similar to Theorem 5.17 is not possible because A_k is not kept fixed for all iterations. A Newton-type method with global convergence proof based on the Newton path is outlined in Section 5.5.

5.4 A Rosenbrock-type example

In order to appreciate the significance of the natural level function let us consider the following example due to Bock [24] and the discussion therein. We use the Newton method to find a zero of the function

$$F(z) = \begin{pmatrix} z_1 \\ 50z_2 + (z_1 - 50)^2/4 \end{pmatrix}$$

with starting guess $z^0 = (50, 1)$ and solution $z^* = (0, -12.5)$ (compare Figure 5.1). We observe that the classical level set (contained within the dashed curve) is shaped like a bent ellipse. The excentricity of the ellipse is due to the condition number of $J(z^0)$, which is $\mathrm{cond}_2(J(z^0)) = 50$. The ellipse is bent because of the mild nonlinearity in the second component of F. A short computation yields $\omega \leq 0.01$.

We further observe that the direction of steepest descent for the classical level function $T(z|\mathbb{I})$ and for the natural level function $T(z|J(z^0)^{-1})$ describe an angle of 87.7 degrees. Thus the Newton increment, which coincides with the direction of steepest descent for the natural level function (see Section 5.3.3), is almost parallel to the tangent on the classical level set. Consequently only heavily damped Newton steps lead to a decrease in the classical level function. We obtain optimal descent in $T(z|\mathbb{I})$ with a stepsize of $\alpha_0 \approx 0.077$, although the solution z^* can be reached with two iterations of a full step Newton method. This behavior illustrates how the requirement of descent in the classical level function impedes fast convergence within the domain of local contraction.

In real-world problems the conditioning of $J(z^*)$ is typically a few orders of magnitude higher than 50, leading to even narrower valleys in the classical level function. Additionally, nonlinear and highly nonlinear problems with larger ω give rise to higher curvature of these valleys, rendering the requirement of descent in the classical level function completely inappropriate.

Especially in the light of inexact Newton methods as we describe in Section 5.6, the use of natural level functions $T(z|J(z^k)^{-1})$ is paramount: Even relatively small perturbations of the exact Newton increment can result in the inexact Newton increment being a direction of ascent in $T(z|\mathbb{I})$.

5.5 The Restrictive Monotonicity Test

Bock et al. [26] have developed a damping strategy called the RMT which can be interpreted as a step size control for integration of the Davidenko differential equation (5.6) with the explicit Euler method. The Euler method can be extended by a number of so called *back projection* steps which diminish the distance of the iterates to the Newton path emanating from z^0. The quantities involved in the first back projection must anyway be computed to control the step size. Numerical experience seems to suggest that more than one back projection step does not improve convergence considerably and should thus be avoided in all known cases. However, repeated back projection steps provide the theoretical benefit of making a proof of global convergence of the RMT possible. In particular, the RMT does not lead to iteration cycles on the notorious example by Ascher and Osborne [7] in contrast to the NMT.

5.6 Natural Monotonicity for LISA-Newton methods

In this section we give a detailed description of an affine covariant globalization strategy for a Newton-type method based on iterative linear algebra. The linear solver must supply error estimates in the variable space. This strategy is described in Deuflhard [44, Section 3.3.4] for the linear solvers CGNE (see, e.g., Saad [135]) and GBIT (due to Deuflhard et al. [45]). We describe a LISA within the Newton-type framework in Section 5.6.2. We have also developed a suitable preconditioner for the problem class (3.3) which we present in Chapter 6.

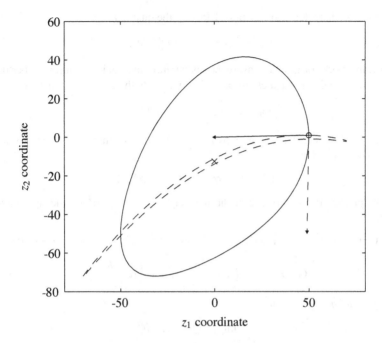

Figure 5.1: Rosenbrock-type example (adapted from Bock [24]): The Newton increment (solid arrow) is almost perpendicular to the direction of steepest descent for the classical level function $T(z|\mathbb{I})$ (dashed arrow) in the initial iterate (marked by $\circ$). Only heavily damped Newton steps lead to descent in the classical level function due to the narrow and bent classical level set (contained within the dashed curve). In contrast, the Newton step is completely contained within the more circle-shaped natural level set (contained within the solid curve) corresponding to the natural level function $T(z|J(z^0)^{-1})$. Within two full Newton steps, the solution (marked by $\times$) is reached.

5.6.1 Estimates for natural monotonicity

Let Δz^k denote the exact Newton step and δz^k the inexact Newton step obtained from LISA. Furthermore, define the residual

$$r^k := J(z^k)(\delta z^k - \Delta z^k).$$

We want to characterize the error of LISA by the quantity

$$\delta_k := \left\| \delta z^k - \Delta z^k \right\| / \left\| \delta z^k \right\|.$$

The framework for global convergence is natural monotonicity subject to perturbations due to the inexactness of δz^k. First, we study the contraction factors

$$\Theta_k(\alpha) = \left\| \overline{\Delta z}^{k+1}(\alpha) \right\| / \left\| \Delta z^k \right\|$$

in terms of the exact Newton steps Δz^k and the exact Simplified Newton steps $\overline{\Delta z}^{k+1}(\alpha)$ defined via

$$J(z^k)\overline{\Delta z}^{k+1}(\alpha) = -F(z^k + \alpha \delta z^k).$$

We emphasize the occurrence of the inexact Newton step δz^k on the right hand side.

Lemma 5.18. *Let $\delta_k < \frac{1}{2}$ and define $h_k^\delta := \widehat{\omega} \left\| \delta z^k \right\|$. Then we obtain the estimate*

$$\Theta_k(\alpha) \leq 1 - \left(1 - \frac{\delta_k}{1 - \delta_k}\right) \alpha + \frac{1}{2}\alpha^2 \frac{h_k^\delta}{1 - \delta_k}. \tag{5.7}$$

The optimal damping factor is

$$\overline{\alpha}_k = \min(1, (1 - 2\delta_k)/h_k^\delta).$$

If we implicitly define ρ via

$$\delta_k = \frac{\rho}{2}\alpha h_k^\delta \tag{5.8}$$

and assume that $\rho \leq 1$ we obtain the optimal damping factor

$$\overline{\alpha}_k = \min(1, 1/\left((1 + \rho)h_k^\delta\right)). \tag{5.9}$$

Proof. See Deuflhard [44, Lemma 3.17]. □

This test can and should not be directly implemented because the computation of the constant h_k^δ and the exact Simplified Newton step $\overline{\Delta z}^{k+1}$ is prohibitively expensive.

An appropriate replacement for the nonrealizable Θ_k is the *inexact Newton path* $\widetilde{z}(\alpha), \alpha \in [0, 1]$, which we can implicitly define via

$$F(\widetilde{z}(\alpha)) = (1 - \alpha)F(z^k) + \alpha r^k.$$

We immediately observe that $\widetilde{z}(0) = z^k, \dot{\widetilde{z}}(0) = \delta z^k$, but $\widetilde{z}(1) \neq z^*$ if $r^k \neq 0$. We can now define an exact Simplified Newton step on the perturbed residual via

$$J(z^k)\widetilde{\Delta z}^{k+1} = -F(z^k + \alpha \delta z^k) + r^k. \tag{5.10}$$

Lemma 5.19. *With the current notation and definitions we obtain the estimate*

$$\left\|\widetilde{\Delta z}^{k+1} - (1-\alpha)\delta z^k\right\| \le \tfrac{1}{2}\alpha^2 h_k^\delta \left\|\delta z^k\right\|.$$

Proof. See Deuflhard [44, Lemma 3.18]. □

For computational efficiency reasons we must also refrain from solving equation (5.10) exactly. Instead we use LISA which introduces another residual error, denoted by $\widetilde{r}_i^{k+1}$ and defined for each *inner iteration* (LISA iteration) i. Then we can define an i-dependent inexact Simplified Newton step via

$$J(z^k)\widetilde{\delta z}_i^{k+1} = (-F(z^k + \alpha\delta z^k) + r^k) + \widetilde{r}_i^{k+1}.$$

As in the above formula, we need to keep the dependence of $\widetilde{\delta z}_i^{k+1}$ on α^k in mind but drop it in the notation for the sake of brevity. It is now paramount for the efficiency of the Newton-type method to balance the accuracies of the inner iterations with the nonlinearity of the problem.

Lemma 5.19 suggests to use a so-called *cross-over of initial values* [44] for LISA according to

$$\widetilde{\delta z}_0^{k+1} = (1-\alpha)\delta z^k, \quad \delta z_0^k = \widetilde{\delta z}^k, \tag{5.11}$$

which predict the solution to first order in α.

We substitute the nonrealizable contraction factor Θ_k now by

$$\widetilde{\Theta}_k = \left\|\widetilde{\delta z}^{k+1}\right\| / \left\|\delta z^k\right\|,$$

which can be computed efficiently. The following lemma characterizes the dependence of $\widetilde{\Theta}_k$ on α.

Lemma 5.20. *Assume that LISA for equation (5.10) with initial value crossing (5.11) has been iterated until*

$$\widetilde{\rho}_i = \frac{\left\|\widetilde{\Delta z}^{k+1} - \widetilde{\delta z}_i^{k+1}\right\|}{\left\|\widetilde{\Delta z}^{k+1} - \widetilde{\delta z}_0^{k+1}\right\|} < 1. \tag{5.12}$$

Then we obtain the estimate

$$\left\|\widetilde{\delta z}_i^{k+1} - (1-\alpha)\delta z^k\right\| \le \frac{1+\widetilde{\rho}_i}{2}\alpha^2 h_k^\delta \left\|\delta z^k\right\|.$$

Proof. The proof for the LISA case is the same as for the GBIT case, see Deuflhard [44, Lemma 3.20]. □

The quantity $\widetilde{\rho}_i$, however, cannot be evaluated directly because we must not compute $\widetilde{\Delta z}^{k+1}$ exactly for efficiency reasons. Instead we define the computable estimate

$$
\overline{\rho}_i = \frac{\left\| \widetilde{\Delta z}^{k+1} - \widetilde{\delta z}_i^{k+1} \right\|}{\left\| \widetilde{\delta z}_i^{k+1} - \widetilde{\delta z}_0^{k+1} \right\|} \approx \frac{\widetilde{\varepsilon}_i}{\left\| \widetilde{\delta z}_i^{k+1} - \widetilde{\delta z}_0^{k+1} \right\|}, \tag{5.13}
$$

where $\widetilde{\varepsilon}_i$ is an estimate for the error of LISA, see Section 5.6.2.

Lemma 5.21. *With the notation and assumptions of Lemma 5.20 we have*

$$
\left\| \widetilde{\Delta z}^{k+1} - \widetilde{\delta z}_i^{k+1} \right\| \leq \overline{\rho}_i (1 + \widetilde{\rho}_i) \left\| \widetilde{\Delta z}^{k+1} - \widetilde{\delta z}_0^{k+1} \right\|.
$$

Proof. We follow Deuflhard [44, Lemma 3.20 for GBIT and below]: The application of the triangle inequality and assumptions (5.12) and (5.11) yield

$$
\left\| \widetilde{\delta z}_i^{k+1} - (1 - \alpha)\delta z^k \right\| \leq \left\| \widetilde{\Delta z}^{k+1} - (1 - \alpha)\delta z^k \right\| + \left\| \widetilde{\delta z}_i^{k+1} - \widetilde{\Delta z}^{k+1} \right\|
$$

$$
= (1 + \widetilde{\rho}_i) \left\| \widetilde{\Delta z}^{k+1} - (1 - \alpha)\delta z^k \right\|.
$$

Using definition (5.13) on the left hand side then delivers the assertion. □

An immediate consequence of Lemma 5.21 is the inequality

$$
\widetilde{\rho}_i \leq \overline{\rho}_i (1 + \widetilde{\rho}_i). \tag{5.14}
$$

Deuflhard [44] proposes to base the estimation of $\widetilde{\rho}_i$ on equating the left and right hand sides of inequality (5.14) to obtain

$$
\overline{\rho}_i = \widetilde{\rho}_i / (1 + \widetilde{\rho}_i), \quad \text{or} \quad \widetilde{\rho}_i = \overline{\rho}_i / (1 - \overline{\rho}_i) \text{ for } \overline{\rho}_i < 1. \tag{5.15}
$$

Then accuracy control for the inner iterations can be based on the termination condition

$$
\widetilde{\rho}_i \leq \widetilde{\rho}_{\max} \quad \text{with} \quad \widetilde{\rho}_{\max} \leq \tfrac{1}{4},
$$

or, following (5.15),

$$
\overline{\rho}_i \leq \overline{\rho}_{\max} \quad \text{with} \quad \overline{\rho}_{\max} \leq \tfrac{1}{3}.
$$

We feel urged to remark that this is heuristic insofar as from inequality (5.14) we can only conclude

$$
\overline{\rho}_i \geq \widetilde{\rho}_i / (1 + \widetilde{\rho}_i) \quad \text{(and not ``$\leq$'')}.
$$

The optimal damping factor α_k from Lemma 5.18 depends on the unknown $h_k^\delta = \widehat{\omega} \left\| \delta z^k \right\|$ which must be approximated. Using the [.] notation (see Remark 5.8) we approximate h_k^δ with a so-called *Kantorovich estimate*

$$[h_k^\delta] = [\widehat{\omega}] \left\| \delta z^k \right\| \leq h_k^\delta,$$

which leads via equation (5.8) to a computable estimate of the optimal step size

$$[\overline{\alpha}_k] = \min(1, (1 - 2\delta_k)/[h_k^\delta]) = \min\left(1, 1/\left((1 + \rho)[h_k^\delta]\right)\right).$$

Based on Lemma 5.20 we obtain an a-posteriori Kantorovich estimate

$$[h_k^\delta]_i = \frac{2 \left\| \widetilde{\delta z}_i^{k+1} - \widetilde{\delta z}_0^{k+1} \right\|}{(1 + \widetilde{\rho}_i)\alpha^2 \left\| \delta z^k \right\|} = \frac{2 \left\| \widetilde{\delta z}_i^{k+1} - (1 - \alpha)\delta z^k \right\|}{(1 + \widetilde{\rho}_i)\alpha^2 \left\| \delta z^k \right\|} \leq h_k^\delta.$$

Replacing $\widetilde{\rho}_i$ by $\overline{\rho}_i$ yields a computable a-posteriori Kantorovich estimate

$$[h_k^\delta]_i = \frac{2(1 - \overline{\rho}_i) \left\| \widetilde{\delta z}_i^{k+1} - \widetilde{\delta z}_0^{k+1} \right\|}{\alpha^2 \left\| \delta z^k \right\|} = \frac{2(1 - \overline{\rho}_i) \left\| \widetilde{\delta z}_i^{k+1} - (1 - \alpha)\delta z^k \right\|}{\alpha^2 \left\| \delta z^k \right\|} \leq h_k^\delta. \tag{5.16}$$

From the definition of $[h_k^\delta]$ we can also derive a computable a-priori Kantorovich estimate

$$[h_{k+1}^\delta] = \frac{\left\| \delta z^{k+1} \right\|}{\left\| \delta z^k \right\|} [h_k^\delta]_*, \tag{5.17}$$

where $[h_k^\delta]_*$ denotes the Kantorovich estimate after the last inner iteration.

The following bit counting lemma finally supplies bounds for the exact and inexact contraction factors.

Lemma 5.22. *Let an inexact Newton method with step sizes $\alpha = [\overline{\alpha}_k]$ be realized. Assume that the leading binary digit of $[h_k^\delta]$ is correct, i.e.,*

$$0 \leq h_k^\delta - [h_k^\delta] < \sigma \max(1/(1 + \rho), [h_k^\delta]) \quad \text{for some } \sigma < 1.$$

Then the exact natural contraction factor satisfies

$$\Theta_k = \frac{\left\| \overline{\Delta z}^{k+1} \right\|}{\left\| \Delta z^k \right\|} < 1 - \frac{1 - \sigma(1 + 2\rho)}{2 + \rho(1 - \sigma)} \alpha.$$

The inexact natural contraction factor is bounded by

$$\widetilde{\Theta}_k = \frac{\widetilde{\delta z}^{k+1}}{\left\| \delta z^k \right\|} < 1 - \left(1 - \frac{1}{2} \frac{(1 + \widetilde{\rho})(1 + \sigma)}{1 + \rho}\right) \alpha. \tag{5.18}$$

Proof. We recall bound (5.7) in the form

$$\Theta_k \le 1 - \left(1 - \frac{\delta_k}{1-\delta_k} - \frac{\alpha h_k^\delta}{2(1-\delta_k)}\right)\alpha. \tag{5.19}$$

We now find a bound for αh_k^δ with optimal realizable step size

$$\alpha = [\overline{\alpha}_k] = \min\left(1, 1/\left((1+\rho)[h_k^\delta]\right)\right).$$

If $\alpha = 1$ we obtain $[h_k^\delta] \le 1/(1+\rho)$ and thus

$$\alpha h_k^\delta \le [h_k^\delta] + \sigma \max(1/(1+\rho), [h_k^\delta]) \le \frac{1+\sigma}{1+\rho}.$$

If $\alpha < 1$ we obtain the same bound

$$\begin{aligned}
\alpha h_k^\delta &= \frac{h_k^\delta}{(1+\rho)[h_k^\delta]} < \frac{[h_k^\delta] + \sigma \max(1/(1+\rho), [h_k^\delta])}{(1+\rho)[h_k^\delta]} \\
&= \frac{1 + \sigma \max(\alpha, 1)}{1+\rho} = \frac{1+\sigma}{1+\rho}.
\end{aligned}$$

Therefore δ_k can be bounded in ρ and σ according to

$$\delta_k = \frac{\rho}{2}\alpha h_k^\delta \le \frac{\rho(1+\sigma)}{2(1+\rho)},$$

which in turn yields for the factor in parentheses in equation (5.19)

$$\begin{aligned}
1 - \frac{\delta_k}{1-\delta_k} - \frac{\alpha h_k^\delta}{2(1-\delta_k)} &\ge 1 - \frac{\frac{\rho(1+\sigma)}{2(1+\rho)}}{1 - \frac{\rho(1+\sigma)}{2(1+\rho)}} - \frac{\frac{1+\sigma}{1+\rho}}{2\left(1 - \frac{\rho(1+\sigma)}{2(1+\rho)}\right)} \\
&= 1 - \frac{\rho(1+\sigma) + (1+\sigma)}{2(1+\rho) - \rho(1+\sigma)} = \frac{2 + \rho(1-\sigma) - \rho(1+\sigma) - 1 - \sigma}{2 + \rho(1-\sigma)} \\
&= \frac{1 - \sigma(1+2\rho)}{2 + \rho(1-\sigma)},
\end{aligned}$$

which proves the first assertion. For the inexact natural contraction factor we use the triangle inequality in combination with Lemma 5.20 to obtain

$$\begin{aligned}
\widetilde{\Theta}_k &= \left\|\widetilde{\delta z}^{k+1}\right\| / \left\|\delta z^k\right\| \le \left((1-\alpha)\left\|\delta z^k\right\| + \left\|\widetilde{\delta z}^{k+1} - (1-\alpha)\delta z^k\right\|\right) / \left\|\delta z^k\right\| \\
&\le 1 - \alpha + \frac{1+\widetilde{\rho}}{2}\alpha^2 h_k^\delta \le 1 - \left(1 - \frac{1}{2}\frac{(1+\widetilde{\rho})(1+\sigma)}{1+\rho}\right)\alpha,
\end{aligned}$$

which shows the second assertion. $\square$

Remark 5.23. With the additional assumption that $\sigma < 1/(1+2\rho)$ we obtain $\Theta_k < 1$.

Based on Lemma 5.22 we can now formulate an inexact NMT. To this end we substitute the bound (5.18) by the computable inexact NMT

$$\widetilde{\Theta}_k = \frac{\left\| \widetilde{\delta z}^{k+1} \right\|}{\left\| \delta z^k \right\|} < 1 - \frac{\rho - \widetilde{\rho}}{1+\rho} \alpha, \tag{5.20}$$

by replacing σ with its upper bound $\sigma = 1$. In order to be a meaningful test we additionally require $\widetilde{\rho} < \rho$. The inexact NMT (5.20) becomes computable if we select a $\rho < 1$ and further impose for the relative error of the inner LISA that

$$\delta_k \leq \frac{\rho}{2} \alpha [h_k^\delta] \leq \frac{\rho(1+\sigma)}{2(1+\rho)}$$

which is substituted by the computable condition

$$\delta_k \leq \frac{\rho}{2(1+\rho)} =: \overline{\delta} \leq \frac{1}{4}$$

for the case of $\alpha < 1$.

If in the course of the computation the inexact NMT is not satisfied, we reduce the step size on the basis of the a-posteriori Kantorovich estimate (5.16), denoted by $[h_k^\delta]_*$, according to

$$\alpha_k^{\text{new}} := \max(\min(1/\left((1+\rho)[h_k^\delta]_*\right), \alpha/2), \alpha_{\text{maxred}} \alpha)\Big|_{\alpha=\alpha_k^{\text{old}}}.$$

Taking the min and max is necessary to safeguard the stepsize adaption against too cautious (reduction by at least a factor of two) and too aggressive changes (reduction by at most a factor of $\alpha_{\text{maxred}} \approx 0.1$). Especially aggressive reduction must be safe-guarded because the computation of the Kantorovich estimate $[h_k^\delta]$ in equation (5.16) is inflicted with a cancellation error which is then amplified by $1/\alpha^2$. Furthermore, the cancellation error gets worse for smaller α because then $\widetilde{\delta z}^{k+1}$ is closer and closer to $(1-\alpha)\delta z^k$ as a consequence of Lemma 5.20.

For the initial choice for α_k we recede to the a-priori Kantorovich estimate (5.17) via

$$\alpha_{k+1} = \min(1, 1/\left((1+\rho)[h_k^\delta]\right))$$

The initial step size α_0 has to be supplied by the user. As a heuristic one can choose $\alpha_0 = 1, 0.01, 0.0001$ for mildly nonlinear, nonlinear, and highly nonlinear problems, respectively.

5.6.2 A Linear Iterative Splitting Approach

The goal of this section is to characterize the convergence and to give error estimates for LISA. Furthermore we address the connection of the convergence rate of LISA with the asymptotic convergence of the LISA-Newton method. To this end let $\hat{J}, \hat{M} \in \mathbb{R}^{N \times N}$ and $\hat{F} \in \mathbb{R}^N$. We approximate $\zeta \in \mathbb{R}^N$ which satisfies

$$\hat{J}\zeta = -\hat{F}$$

via the iteration

$$\zeta_{i+1} = \zeta_i - \hat{M}\left(\hat{J}\zeta_i + \hat{F}\right) = (\mathbb{I} - \hat{M}\hat{J})\zeta_i - \hat{M}\hat{F}. \tag{5.21}$$

The iteration is formally based on the splitting

$$\hat{J} = \hat{M}^{-1} - \widehat{\Delta J}.$$

We have been using this setting in Section 5.6.1 with $\hat{J} = J(z^k)$ and $\hat{F} = F(z^k)$ or $\hat{F} = F(z^k + \alpha_k \delta z^k)$ to approximate $\zeta = \Delta z^k$ or $\zeta = \widetilde{\Delta z}^{k+1}$, respectively. The matrix $\hat{M}$ is a preconditioner which can be used in a truncated Neumann series to describe the approximated inverse $M(z)$. We address this issue later in Lemma 5.27.

5.6.2.1 Affine invariant convergence of LISA

Lemma 5.24. *Let* $A, B \in \mathrm{GL}(N)$ *yield transformations of* $\hat{F}, \hat{J},$ *and* $\hat{M}$ *which satisfy*

$$\widetilde{F} = A\hat{F}, \quad \widetilde{J} = A\hat{J}B, \quad \widetilde{M} = B^{-1}\hat{M}A^{-1}.$$

Then LISA is affine invariant under A and B.

Proof. Assume $\widetilde{\zeta}_i = B^{-1}\zeta_i$. Then we have

$$\widetilde{\zeta}_{i+1} = (\mathbb{I} - \widetilde{M}\widetilde{J})\widetilde{\zeta}_i - \widetilde{M}\widetilde{F} = \left(\mathbb{I} - B^{-1}\hat{M}A^{-1}A\hat{J}B\right)B^{-1}\zeta_i - B^{-1}\hat{M}A^{-1}A\hat{F}$$
$$= B^{-1}\left[\left(\mathbb{I} - \hat{M}\hat{J}\right)\zeta_i - \hat{M}\hat{F}\right] = B^{-1}\zeta_{i+1}.$$

Induction yields the assertion. □

Corollary 5.25. *A full-step LISA-Newton method is affine invariant under transformations* $A, B \in \mathrm{GL}(N)$ *with*

$$\widetilde{F}(\widetilde{z}) = A\hat{F}(B\widetilde{z})$$

if the matrix function $\hat{M}(z)$ *satisfies*

$$\widetilde{M}(\widetilde{z}) = B^{-1}\hat{M}(B\widetilde{z})A^{-1}.$$

Proof. Lemmata 5.9 and 5.24. □

The Newton-Picard preconditioners in Chapter 6 satisfy this relation at least partially which leads to scaling invariance of the Newton-Picard LISA-Newton method (see Chapter 6).

The convergence requirements of LISA are described by the following theorem:

Theorem 5.26. *Let*

$$\kappa_{\text{lin}} := \sigma_r(\mathbb{I} - \hat{M}\hat{J}).$$

If $\kappa_{\text{lin}} < 1$ then $\hat{M}$ and $\hat{J}$ are invertible and LISA (5.21) converges for all $\hat{F}, \zeta_0 \in \mathbb{R}^N$. Conversely, if LISA converges for all $\hat{F}, \zeta_0 \in \mathbb{R}^N$, then $\kappa_{\text{lin}} < 1$. The asymptotic linear convergence factor is given by κ_{lin}.

Proof. See Saad [135, Theorem 4.1]. □

5.6.2.2 Connection between linear and nonlinear convergence

We now investigate the connection of the preconditioner $\hat{M}(z)$ with the approximated inverse $M(z)$ of the Local Contraction Theorem 5.5. The results have already been given by Ortega and Rheinboldt [122, Theorem 10.3.1]. We translate them into the framework of the Local Contraction Theorem 5.5.

Lemma 5.27. *Let $\zeta_0 = 0$ and $l \geq 1$. Then the l-th iterate of LISA is given by the truncated Neumann series for $\hat{M}\hat{J}$ according to*

$$\zeta_l = - \left[\sum_{i=0}^{l-1} (\mathbb{I} - \hat{M}\hat{J})^i \right] \hat{M}\hat{F}.$$

Proof. Let $l \in \mathbb{N}$ and assume that the assertion holds for ζ_l. Then we obtain

$$\zeta_{l+1} = (\mathbb{I} - \hat{M}\hat{J})\zeta_l - \hat{M}\hat{F} = - \left[\left(\sum_{i=0}^{l-1} (\mathbb{I} - \hat{M}\hat{J})^{i+1} \right) + \mathbb{I} \right] \hat{M}\hat{F}$$

$$= - \left[\sum_{i=0}^{l} (\mathbb{I} - \hat{M}\hat{J})^i \right] \hat{M}\hat{F}.$$

For $l = 1$ we have $\zeta_1 = -\hat{M}\hat{F}$ and we complete the proof by induction. □

The following lemma shows that M, defined by l LISA steps, is almost the inverse of J.

Lemma 5.28. *Let the approximated inverse be defined according to*

$$M(z) = \left[\sum_{i=0}^{l-1} (\mathbb{I} - \hat{M}(z)J(z))^i \right] \hat{M}(z) = \hat{M}(z) \left[\sum_{i=0}^{l-1} (\mathbb{I} - J(z)\hat{M}(z))^i \right].$$

Then it holds that

$$M(z)J(z) = \mathbb{I} - (\mathbb{I} - \hat{M}(z)J(z))^l,$$
$$J(z)M(z) = \mathbb{I} - (\mathbb{I} - J(z)\hat{M}(z))^l.$$

Proof. With the abbreviation $A := \mathbb{I} - \hat{M}(z)J(z)$ we obtain the first assertion

$$M(z)J(z) = \left[\sum_{i=0}^{l-1} A^i \right] (\mathbb{I} - A) = \sum_{i=0}^{l-1} A^i - \sum_{i=1}^{l} A^i = \mathbb{I} - (\mathbb{I} - \hat{M}(z)J(z))^l.$$

The second assertion follows with the same argument. □

Theorem 5.29. *Let $z^* \in D$ satisfy $\hat{M}(z^*)F(z^*) = 0$. For the LISA-Newton method with continuous preconditioner $\hat{M}(z)$ and l steps of LISA with starting guess $\zeta_0 = 0$, the following two assertions are equivalent:*

i) The LISA at z^* converges for all starting guesses and right hand sides, i.e.,

$$\sigma_r(\mathbb{I} - \hat{M}(z^*)J(z^*)) \le \kappa_{\text{lin}} < 1.$$

ii) The matrices $\hat{M}(z^*)$ and $J(z^*)$ are invertible and for every $\varepsilon > 0$ there exists a norm $\|.\|_*$ and a neighborhood U of z^* such that the κ-condition 5.3 for $M(z)$ in U based on $\|.\|_*$ is satisfied with

$$\kappa \le \kappa_{\text{lin}}^l + \varepsilon \quad (\text{where } \kappa_{\text{lin}} < 1).$$

Proof. i) $\Rightarrow$ ii): By virtue of Theorem 5.26 it holds that

$$\sigma_r(\mathbb{I} - \hat{M}(z^*)J(z^*)) \le \kappa_{\text{lin}} < 1$$

and that $\hat{M}(z^*)$ and $J(z^*)$ are invertible. Recall that we have

$$\mathbb{I} - M(z^*)J(z^*) = (\mathbb{I} - \hat{M}(z^*)J(z^*))^l$$

by Lemma 5.28. Let $\varepsilon > 0$. Then the Hirsch Theorem delivers a norm $\|.\|_*$ such that

$$\|\mathbb{I} - M(z^*)J(z^*)\|_* \le \sigma_r(\mathbb{I} - \hat{M}(z^*)J(z^*))^l + \varepsilon/2 \le \kappa_{\text{lin}}^l + \varepsilon/2.$$

By continuity of $F, J, \hat{M}, \|.\|_*$, and the inverse we obtain the existence of a neighborhood U of z^* such that all $z \in U$ satisfy

$$\|M(z - M(z)F(z))(\mathbb{I} - J(z)M(z))F(z)\|_*$$
$$\leq \left\|M(z - M(z)F(z))(M^{-1}(z) - J(z))\right\|_* \|M(z)F(z)\|_*$$
$$\leq (\kappa_{\text{lin}}^l + \varepsilon) \|M(z)F(z)\|_*$$

because

$$M(z^* - M(z^*)F(z^*))(M^{-1}(z^*) - J(z^*)) = \mathbb{I} - M(z^*)J(z^*).$$

Comparison with the κ-condition 5.3 yields

$$\kappa \leq \kappa_{\text{lin}}^l + \varepsilon.$$

ii) $\Rightarrow$ i): Let $\hat{z} \in \mathbb{R}^N$ be a nonzero vector, small enough such that

$$z(t) := z^* + t\hat{z} \in U, \quad t \in (0, 1]. \tag{5.22}$$

Because $M(z^*)$ is invertible we obtain $F(z^*) = 0$ and write $F(z(t))$ in the form

$$F(z(t)) = F(z^* + t\hat{z}) - F(z^*) = t \int_0^1 J(z^* + \tau t \hat{z}) \hat{z} d\tau,$$

which leads to

$$z'(t) - z(t) := -M(z(t))F(z(t)) = -tM(z(t)) \int_0^1 J(z(\tau t)) d\tau \hat{z}.$$

From the κ-condition 5.3 we infer the inequality

$$t \left\| M(z'(t)) \left[M(z(t))^{-1} - J(z(t)) \right] M(z(t)) \int_0^1 J(z(\tau t)) d\tau \hat{z} \right\|_*$$
$$\leq t\kappa \left\| M(z(t)) \int_0^1 J(z(\tau t)) d\tau \hat{z} \right\|_*.$$

After division by t we take the limit $t \to 0$ and obtain

$$\kappa \|M(z^*)J(z^*)\hat{z}\|_* \geq \|M(z^*)J(z^*) \left[\mathbb{I} - M(z^*)J(z^*)\right] \hat{z}\|_*. \tag{5.23}$$

We now recede to a special choice for $\hat{z}$ based on the real Jordan decomposition

$$\mathbb{I} - M(z^*)J(z^*) = V \begin{pmatrix} J_1 & & \\ & \ddots & \\ & & J_p \end{pmatrix} V^{-1},$$

with an invertible matrix $V \in \mathbb{R}^{N \times N}$ and real Jordan blocks J_i, corresponding to eigenvalues $\lambda_i = a_i + ib_i$, of the form

$$
J_i = \begin{pmatrix} a_i & 1 & & \\ & a_i & \ddots & \\ & & \ddots & 1 \\ & & & a_i \end{pmatrix} \text{ if } b_i = 0 \quad \text{and} \quad J_i = \begin{pmatrix} C_i & \mathbb{I}_2 & & \\ & C_i & \ddots & \\ & & \ddots & \mathbb{I}_2 \\ & & & C_i \end{pmatrix} \text{ if } b_i \neq 0,
$$

where

$$
C_i = \begin{pmatrix} a_i & b_i \\ -b_i & a_i \end{pmatrix}.
$$

We assume that the eigenvalues are sorted to satisfy $|\lambda_i| \geq |\lambda_{i+1}|$. Let V_j denote the j-th column of V. If $b_1 = 0$, we choose $\alpha \neq 0$ small enough such that with $\hat{z} = \alpha V_1$ assumption (5.22) holds. We then obtain

$$
\|M(z^*)J(z^*)[\mathbb{I} - M(z^*)J(z^*)]\hat{z}\| = |\lambda_1| \|M(z^*)J(z^*)\hat{z}\|. \tag{5.24}
$$

If $b_1 \neq 0$, then we choose $\alpha, \beta \in \mathbb{R}$ such that $\hat{z} = \alpha V_1 + \beta V_2 \neq 0$ and (5.22) holds. Due to invertibility of $M(z^*), J(z^*)$, and V, we can define the norm

$$
\left\| \begin{pmatrix} \alpha \\ \beta \end{pmatrix} \right\|_V = \left\| M(z^*)J(z^*)(V_1 \ V_2) \begin{pmatrix} \alpha \\ \beta \end{pmatrix} \right\|_*
$$

and consider the estimate

$$
\|M(z^*)J(z^*)[\mathbb{I} - M(z^*)J(z^*)]\hat{z}\|_*
$$
$$
= \left\| C_1 \begin{pmatrix} \alpha \\ \beta \end{pmatrix} \right\|_V \geq |\lambda_1| \left\| \begin{pmatrix} \alpha \\ \beta \end{pmatrix} \right\|_V = |\lambda_1| \|M(z^*)J(z^*)\hat{z}\|_*. \tag{5.25}
$$

Finally we can combine (5.23), (5.24), and (5.25) in order to establish

$$
\kappa \|M(z^*)J(z^*)\hat{z}\|_* \geq \|M(z^*)J(z^*)[\mathbb{I} - M(z^*)J(z^*)]\hat{z}\|_* \geq |\lambda_1| \|M(z^*)J(z^*)\hat{z}\|_*.
$$

Thus we have

$$
\kappa_{\text{lin}}^l + \varepsilon \geq \kappa \geq |\lambda_1| = \sigma_r(\mathbb{I} - M(z^*)J(z^*)) = \sigma_r(\mathbb{I} - \hat{M}(z^*)J(z^*))^l,
$$

independently of $\|.\|_*$. Letting $\varepsilon \to 0$ yields assertion i). $\square$

Let us halt shortly to discuss the previous results: Far away from the solution the use of LISA allows for adaptive control of the angle between the search direction and the tangent on the Newton path according to Lemma 5.28. Theorem 5.29 guarantees that although a LISA-Newton method with larger number l of inner iterations is numerically more expensive per outer iteration than using $l = 1$, the numerical effort in the vicinity of the solution is asymptotically fully compensated by less outer inexact Newton iterations.

5.6.2.3 Convergence estimates for LISA

To simplify the presentation we denote the LISA iteration matrix by

$$A = \mathbb{I} - \hat{M}\hat{J}.$$

Lemma 5.30. *Assume* $\|A\| \leq \hat{\kappa} < 1$. *Then the following estimates hold:*

$$\|\zeta_{l+1} - \zeta_l\| \leq \hat{\kappa}^l \|\zeta_1 - \zeta_0\|,$$

$$\|\zeta_l - \zeta_\infty\| \leq \frac{\hat{\kappa}^l}{1 - \hat{\kappa}} \|\zeta_1 - \zeta_0\|.$$

Proof. Let $l \geq 1$. The first assertion follows from

$$\|\zeta_{l+1} - \zeta_l\| = \|A(\zeta_l - \zeta_{l-1})\| \leq \left\|A^l\right\| \|\zeta_1 - \zeta_0\| \leq \hat{\kappa}^l \|\zeta_1 - \zeta_0\|.$$

Thus we obtain

$$\|\zeta_l - \zeta_\infty\| \leq \sum_{k=l}^{\infty} \|\zeta_k - \zeta_{k+1}\| \leq \sum_{k=l}^{\infty} \hat{\kappa}^k \|\zeta_1 - \zeta_0\| \leq \frac{\hat{\kappa}^l}{1 - \hat{\kappa}} \|\zeta_1 - \zeta_0\|,$$

which proves the second assertion. □

5.6.2.4 Estimation of $\hat{\kappa}$

In order to make use of Lemma 5.30 we need a computable estimate $[\hat{\kappa}]$ of $\hat{\kappa}$. We present three approaches which are all based on eigenvalue techniques. They differ mainly in the assumptions on the iteration matrix A, in the required numerical effort, and in memory consumption.

For $l = 1, 2, \ldots$ we define

$$\delta\zeta_l = \zeta_l - \zeta_{l-1}.$$

We have already observed in the proof of Lemma 5.30 that

$$\delta\zeta_{l+1} = A\delta\zeta_l = A^l\delta\zeta_1.$$

Thus LISA behaves like a Power Method (see, e.g., Golub and van Loan [61]). The common idea behind all three $\hat{\kappa}$ estimators is to obtain a good estimate for $\sigma_r(A)$ by approximation of a few eigenvalues during LISA. Based on Theorem 5.29 we expect $\sigma_r(A)$ to be a good asymptotic estimator for the norm-dependent $\hat{\kappa}$.

Lemma 5.31 (Rayleigh κ-estimator). *Let A be diagonalizable and the eigenvalues $\mu_i, i = 1, \ldots, N$ be ordered according to*

$$|\mu_1| > |\mu_2| \geq \cdots \geq |\mu_N|$$

with a gap in modulus between the first and second eigenvalue. If furthermore $\delta\zeta_1$ has a component in the direction of the eigenvector corresponding to μ_1 we obtain

$$[\hat{\kappa}]_l := \frac{\delta\zeta_l^T \delta\zeta_{l+1}}{\delta\zeta_l^T \delta\zeta_l} \to \sigma_r(A) \quad for \ l \to \infty.$$

Proof. The proof coincides with the convergence proof for the Power Method. For a discussion of the convergence we refer the reader to Wilkinson [165] and Parlett and Poole [124]. The quotient

$$\frac{\delta\zeta_l^T \delta\zeta_{l+1}}{\delta\zeta_l^T \delta\zeta_l} = \frac{\delta\zeta_l^T A \delta\zeta_l}{\delta\zeta_l^T \delta\zeta_l}.$$

in the assertion is the Rayleigh quotient. □

We observe that only the last iterate needs to be saved in order to evaluate the Rayleigh κ-estimator which can thus be implemented efficiently with low memory requirements. The possibly slow convergence of $\delta\zeta_l$ towards the dominant eigenvector if $|\mu_1|$ is close to $|\mu_2|$ does not pose a problem for the Rayleigh κ-estimator because we are only interested in the eigenvalue, not the corresponding eigenvector. However, the Rayleigh κ-estimator is not suitable in many practical applications because the assumption that A is diagonalizable is often violated.

We have developed the following κ-estimator for general matrices:

Lemma 5.32 (Root κ-estimator). *Let $\sigma_r(A) > 0$ and let $\delta\zeta_1$ have a component in the dominant invariant subspace corresponding to the eigenvalues of A with largest modulus. Then the quotient of roots*

$$[\hat{\kappa}]_{l+1} := \frac{\|\delta\zeta_{l+1}\|^{1/l}}{\|\delta\zeta_1\|^{1/l}}, \quad l \geq 1,$$

yields an asymptotically correct estimate of $\sigma_r(A)$ for $l \to \infty$.

Proof. Matrix submultiplicativity yields the upper bound

$$\frac{\left\|\delta\zeta_{l+1}\right\|^{1/l}}{\left\|\delta\zeta_1\right\|^{1/l}} = \frac{\left\|A^l\delta\zeta_1\right\|^{1/l}}{\left\|\delta\zeta_1\right\|^{1/l}} \leq \frac{\left\|A^l\right\|^{1/l}\left\|\delta\zeta_1\right\|^{1/l}}{\left\|\delta\zeta_1\right\|^{1/l}} = \left\|A^l\right\|^{1/l},$$

which tends to $\sigma_r(A)$ for $l \to \infty$.

We construct a lower bound in three steps: First, we write down a Jordan decomposition

$$A = X\Lambda X^{-1},$$

where $X \in \mathrm{GL}(N)$ and Λ is a block diagonal matrix consisting of m Jordan blocks $J_{p_j}(\lambda_j)$ of sizes $p_j, j = 1, \ldots, m$, corresponding to the eigenvalues λ_j. From the identity

$$A^l = X\Lambda^l X^{-1}$$

we see that the columns of X corresponding to each Jordan block span a cyclic invariant subspace of A^l. There exists a constant $c > 0$ such that

$$\|z\| \geq c\|z\|_X := c\left\|X^{-1}z\right\|_2$$

because all norms on a finite dimensional space are equivalent. With $z := X^{-1}\delta\zeta_1$ we obtain a reduction of the problem to Jordan form

$$\left\|A^l\delta\zeta_1\right\|^{1/l} \geq c^{1/l}\left\|\Lambda^l z\right\|_2^{1/l}. \tag{5.26}$$

Second, we reduce further to one Jordan block via

$$\left\|\Lambda^l z\right\|_2^2 = \sum_{j=1}^{m}\left\|J_{p_j}(\lambda_j)\tilde{z}^j\right\|_2^2 \geq \left\|J_{p_1}(\lambda_1)\tilde{z}^1\right\|_2^2, \tag{5.27}$$

where $\tilde{z}^j$ is the subvector of z corresponding to the Jordan block $J_{p_j}(\lambda_j)$. Without loss of generality we choose the ordering of the Jordan blocks such that $|\lambda^1| = \sigma_r(A)$ and $\tilde{z}^1 \neq 0$ due to the assumption of the lemma.

Third, we investigate one single Jordan block $J_{p_1}(\lambda_1)$. To avoid unnecessary notational clutter we drop the $j = 1$ indices. Let $l \geq p$. Then we obtain

$$
\sigma_{\mathrm{r}}(A)^{-l} \left\| J_p(\lambda)^l \tilde{z} \right\|_2 = |\lambda|^{-l} \left\| \begin{pmatrix} \lambda^l & \binom{l}{1}\lambda^{l-1} & \cdots & \binom{l}{p-1}\lambda^{l-(p-1)} \\ & \ddots & \ddots & \vdots \\ & & \ddots & \binom{l}{1}\lambda^{l-1} \\ & & & \lambda^l \end{pmatrix} \tilde{z} \right\|_2
$$

$$
= \binom{l}{p-1} \left\| \begin{pmatrix} \frac{1}{\binom{l}{p-1}} & \frac{\binom{l}{1}}{\binom{l}{p-1}}\lambda^{-1} & \cdots & \frac{\binom{l}{p-1}}{\binom{l}{p-1}}\lambda^{-(p-1)} \\ & \ddots & \ddots & \vdots \\ & & \ddots & \frac{\binom{l}{1}}{\binom{l}{p-1}}\lambda^{-1} \\ & & & \frac{1}{\binom{l}{p-1}} \end{pmatrix} \tilde{z} \right\|_2
$$

$$
\underbrace{\phantom{\begin{pmatrix} \frac{1}{\binom{l}{p-1}} & \frac{\binom{l}{1}}{\binom{l}{p-1}}\lambda^{-1} \end{pmatrix}}}_{=:\tilde{A}(l)}
$$

$$
\geq \left\| \tilde{A}(l)\tilde{z} \right\|_2.
$$

To estimate the quotients of binomials we assume $k > j$ and obtain

$$
\frac{\binom{l}{j}}{\binom{l}{k}} = \frac{k!(l-k)!}{j!(l-j)!} = \frac{k(k-1)\cdots(k-j+1)}{(l-j)\cdots(l-k+1)},
$$

which tends to zero for $l \to \infty$. This shows that the $(1,p)$ entry of $\tilde{A}(l)$ dominates for large l. Thus $\left\| \tilde{A}(l)\tilde{z} \right\|_2$ converges to

$$
\sigma_{\mathrm{r}}(A)^{1-p} |\tilde{z}_p| > 0.
$$

(If $|\tilde{z}_p| = 0$ we can use the same argument with the last non-vanishing component of $\tilde{z}$.) Consequently we can find $l_0 \in \mathbb{N}$ such that

$$
\sigma_{\mathrm{r}}(A)^{-l} \left\| J_p(\lambda)^l \tilde{z} \right\|_2 \geq \frac{1}{2}\sigma_{\mathrm{r}}(A)^{1-p} |\tilde{z}_p| > 0 \quad \text{for all } l \geq l_0. \tag{5.28}
$$

We now combine equations (5.26), (5.27), and (5.28) and obtain for $l \geq l_0$ that

$$
\frac{\left\| A^l \delta\zeta_1 \right\|^{1/l}}{\left\| \delta\zeta_1 \right\|^{1/l}} \geq \sigma_{\mathrm{r}}(A) \left(\frac{c\sigma_{\mathrm{r}}(A)^{1-p} |\tilde{z}_p|}{2\left\| \delta\zeta_1 \right\|} \right)^{1/l},
$$

which tends to $\sigma_r(A)$ for $l \to \infty$. $\square$

We believe it is helpful at this point to investigate two prototypical classes of non-diagonalizable matrices to appreciate the convergence of LISA from a geometrical point of view. We restrict the discussion to the case $\hat{F} = 0$ because we can then exploit the fact that $\zeta_{l+1} = A\zeta_l$ as well as $\delta\zeta_{l+1} = A\delta\zeta_l$.

Example 3 (Jordan matrices). In the proof of Lemma 5.32 we have already seen that the cyclic invariant subspaces of A are spanned by the columns of X corresponding to each Jordan block. Thus the convergence in the case where A is one Jordan block is prototypical.

Let thus $\lambda \in [0, 1)$ and

$$A = J_N(-\lambda) := \begin{pmatrix} -\lambda & 1 & & \\ & \ddots & \ddots & \\ & & \ddots & 1 \\ & & & -\lambda \end{pmatrix}.$$

We immediately see that $\sigma_r(A) = \lambda < 1$ and so we obtain convergence of the iteration for all starting values ζ_0 and right hand sides $\hat{F}$ by virtue of Theorem 5.26. In particular, we now choose $\zeta_0 = e_N$, the last column of the N-by-N identity matrix. If $\lambda = 0$ we obtain

$$\zeta_l = \begin{cases} e_{N-l} & \text{for } l < N, \\ 0 & \text{for } l \geq N, \end{cases}$$

i.e., ζ_l circulates backwards through all basis vectors e_j and then suddenly drops to zero. Figure 5.2 depicts iterations with varying λ and N. We observe that the Jordan structure leads to non-monotone transient convergence behavior in the first iterations. Only in the diagonalizable case of $N = 1$ is the convergence monotone.

Figure 5.2 also suggests that the Root κ-estimator of Lemma 5.32 can grossly overestimate $\sigma_r(A)$ in a large preasymptotic range of iterations.

Example 4 (Multiple eigenvalues of largest modulus). Consider the N-by-N permutation matrix

$$P = \begin{pmatrix} & 1 & & \\ & & \ddots & \\ & & & 1 \\ 1 & & & \end{pmatrix}.$$

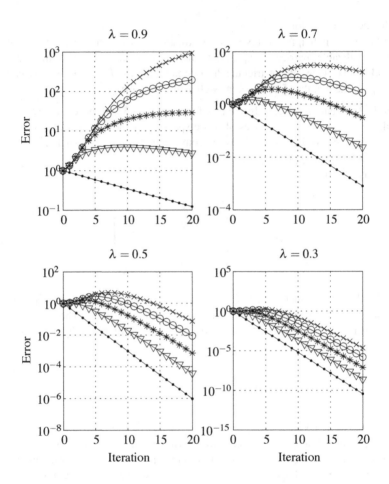

Figure 5.2: The errors of the iterates of LISA in the Euclidean norm $\|.\|_2$ with a Jordan iteration matrix given by Example 3 exhibit non-monotone transient convergence behavior. The subfigures depict different values for λ. The iterations are performed with values for $N = 1, \ldots, 5$, marked by $\bullet, \triangledown, *, \circ, \times$, respectively.

The eigenvalues of P are given by the complex roots of the characteristic polynomial $\lambda^N - 1$. Thus they all satisfy $|\lambda_i| = 1$. Let $X \in \mathrm{GL}(N)$ and $\kappa \in [0, 1)$. Then the matrix

$$A := \kappa X P X^{-1}$$

has all the eigenvalues satisfying $|\kappa\lambda_i| = \kappa = \sigma_r(A)$. Again, Theorem 5.26 yields convergence of LISA for all starting values ζ_0. By virtue of $P^N = \mathbb{I}$ we obtain

$$A^{jN} = (\kappa^N)^j \mathbb{I},$$

which results in monotone N-step convergence. The behavior between the first and N-th step can be non-monotone in an arbitrary norm as displayed in Figure 5.3. If we take instead the X-norm

$$\|z\|_X := \left\| X^{-1}z \right\|_2$$

we immediately obtain monotone convergence by virtue of

$$\|A\zeta\|_X \leq \left(\sup_{\|z\|_X = 1} \|Az\|_X \right) \|\zeta\|_X = \left(\sup_{\|X^{-1}z\|_2 = 1} \kappa \left\| PX^{-1}z \right\|_2 \right) \|\zeta\|_X = \kappa \|\zeta\|_X.$$

In practical computations, however, the construction of a Hirsch-type norm like $\|\cdot\|_X$ is virtually impossible and thus a κ-estimator should be norm-independent.

This leads us to a third approach for the estimation of $\hat{\kappa}$.

Lemma 5.33 (Ritz κ-estimator). *Let $\delta\zeta_1$ have a component in the dominant invariant subspace corresponding to the eigenvalues of A with largest modulus. Define*

$$Z(i,j) = (\delta\zeta_i, \ldots, \delta\zeta_j)$$

and let $R \in \mathbb{R}^{p\times p}$ be an invertible matrix such that $Z(1,p) = QR$ with orthonormal $Q \in \mathbb{R}^{N\times p}$ and maximal $p \leq l$. Then

$$[\hat{\kappa}]_{l+1} := \sigma_r(R^{-T}Z(1,p)^T Z(2,p+1)R^{-1})$$

yields the exact $\sigma_r(A)$ after at most N iterations.

Proof. Consider the Ritz values $\mu_j, j = 1, \ldots, p$ of A on the Krylov space

$$\mathscr{K}_l(A, \delta\zeta_1) := \text{span}(A^0 \delta\zeta_1, \ldots, A^l \delta\zeta_1).$$

The Ritz values solve the following variational eigenvalue problem: Find $v \in \mathscr{K}_l(A, \delta\zeta_1)$ such that

$$w^T(Av - \mu v) = 0, \quad \text{for all } w \in \mathscr{K}_l(A, \delta\zeta_1). \tag{5.29}$$

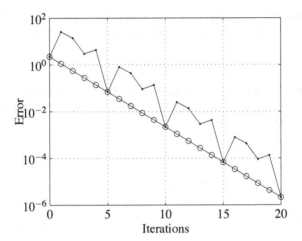

Figure 5.3: The errors of the iterates of LISA with a 5-by-5 matrix $A = \frac{1}{2}XPX^{-1}$ given by Example 4 exhibit non-monotone cyclic convergence behavior in the Euclidean norm $\|.\|_2$ (• marks). The convergence is monotone if measured in the X-norm $\|z\|_X := \|X^{-1}z\|_2$ (∘ marks). We chose the matrix X to be a random symmetric matrix with condition number 100.

Because Q spans an orthonormal basis of $\mathscr{K}_l(A, \delta\zeta_1)$ equation (5.29) is equivalent to the standard eigenvalue problem

$$Q^T(AQ\tilde{v} - \mu Q\tilde{v}) = Q^T AQ\tilde{v} - \mu\tilde{v} = 0.$$

Recall that $Q = Z(1, p)R^{-1}$. Thus we substitute

$$H := Q^T AQ = R^{-T}Z(1, p)^T AZ(1, p)R^{-1} = R^{-T}Z(1, p)^T Z(2, p+1)R^{-1}.$$

Hence, as soon as the dimension of the Krylov space $\mathscr{K}_l(A, \delta\zeta_1) \subseteq \mathbb{R}^N$ becomes stationary when l grows, we obtain $\sigma_r(H) = \sigma_r(A)$. □

The Ritz κ-estimator of Lemma 5.33 yields the most reliable estimates for the spectral radius of A. However, the large memory requirement for storing $Z(1, l)$ is not feasible in practice. Our experience is that a moderate bound on p still provides useful estimates for $\hat{\kappa}$.

Remark 5.34. In the implementation MUSCOP, which we describe in Chapter 11, we explicitly compute Q and R by a QR decomposition. This extra effort is negligible if the matrix vector products with A dominate the overall effort, which is certainly the case in MUSCOP especially on finer spatial grids.

Remark 5.35. We further propose that one should build the matrix Q iteratively, e.g., via an Arnoldi process with upper Hessenberg matrix R (see, e.g., Golub and van Loan [61]). This raises a couple of further questions which would have to be addressed and exceed the scope of this thesis: If an orthonormal basis of the Krylov space is anyway available, is a different solver for the linear systems more appropriate? GMRES by Saad and Schultz [136], e.g., is explicitly built on an Arnoldi process but lacks the property of affine invariance in the residual space and an error criterion in the variable space. Furthermore, a connection between the nonlinear κ and a descriptive constant for convergence of the linear solver like in Theorem 5.29 should be investigated.

5.6.2.5 Adaptive κ improvement

Based on the κ-estimators from Section 5.6.2.4 we can adaptively control the quality of the preconditioner $M(z)$. The procedure is as follows: Let $\kappa_{max} < 1$ and an integer i_{pre} be given. If in the i-th LISA iteration

$$i > i_{pre} \quad \text{and} \quad [\hat{\kappa}]_i > \kappa_{max} \qquad (5.30)$$

then we need to improve the quality of $M(x)$ to decrease κ. The integer i_{pre} is a safeguard to discard preasymptotic estimates of $\hat{\kappa}$ which have not come close to the actual spectral radius of the iteration matrix yet. In our numerical experience with the applications that we present in Part III, $\kappa_{max} = \sqrt{1/2}$ and $i_{pre} = 8$ produce reasonable results.

Depending on the type of preconditioner $M(z)$ the improvement can consist of different strategies: In an adaptive Simplified Newton Method, e.g., we keep $M(z)$ constant until condition (5.30) is satisfied which triggers a new evaluation of M at the current iterate z^k. In Chapter 6 we describe a two-grid preconditioner $M(z)$ which can be improved by refinement of the coarse grid if condition (5.30) holds.

5.6.3 GINKO Algorithm

We distill the algorithmic ingredients for the presented *Global inexact Newton method with κ and ω monitoring (GINKO)* into concise form in Algorithm 1.

5.7 Inequality constrained optimization problems

We have developed an approach how inequality constrained optimization problems can be treated on the basis of an NMT LISA-Newton method (see Section 5.6). We especially focus on the direct use of the GINKO Algorithm 1 for the solution of NLP problem (4.1) which we have formulated in Chapter 4 as

$$\underset{x\in\mathbb{R}^n}{\text{minimize}}\, f(x) \quad \text{s.t.} \quad g_i(x) = 0, i \in \mathscr{E}, \quad g_i(x) \geq 0, i \in \mathscr{I},$$

with $\mathscr{E} \cup \mathscr{I} = \{1,\dots,m\} =: \overline{m}$.

The quintessence of our approach is to formulate the stationarity and primal feasibility condition of the KKT conditions (4.2) in a function F and ensure that dual feasibility and complementarity hold in the solution via suitable choice of M.

We treat the case with exact derivatives in Section 5.7.1 and the extension to inexact derivatives in a LISA-Newton method in Section 5.7.2.

5.7.1 SQP with exact derivatives

SQP is a collective term for certain methods that find critical points of NLP problems. A critical point is a pair $z = (x,y) \in \mathbb{R}^{n+m}$ of primal and dual variables which satisfies the KKT conditions (see Theorem 4.9). SQP methods approximate critical points via sequential solution of QPs which stem from some (approximated) linearization of the NLP around the current iterate. Various variants exist which differ mostly in the way how QP subproblems are formulated and which globalization strategy is used. For an introduction see, e.g., Nocedal and Wright [121].

We now present our novel SQP approach. Let $N = n+m$ and the function $F : \mathbb{R}^N \to \mathbb{R}^N$ be defined according to

$$F(z) = \begin{pmatrix} F_1(z) \\ F_2(z) \end{pmatrix} := \begin{pmatrix} \nabla_x \mathscr{L}(x,y) \\ g(x) \end{pmatrix} \tag{5.31}$$

with Jacobian

$$J(z) = \begin{pmatrix} J_1(z) & -J_2(z)^{\mathrm{T}} \\ J_2(z) & 0 \end{pmatrix} = \begin{pmatrix} \nabla^2_{xx}\mathscr{L}(z) & -\nabla g(x) \\ \nabla g(x)^{\mathrm{T}} & 0 \end{pmatrix}.$$

We observe that J is in general singular because g is not restricted to only active constraints. For instance if g contains upper and lower bounds on a variable then the corresponding two columns in $\nabla g(x)$ are linearly dependent for all $z \in \mathbb{R}^N$. We shall see later that J is invertible on a suitably defined subspace (see Remark 5.38).

Now we generalize the use of an approximated inverse *matrix* $M(z)$ in the step computation to a *nonlinear* function $J^{\oplus} : \mathbb{R}^{N+N} \to \mathbb{R}^N$ to compute

$$\Delta z = J^{\oplus}(z, -\hat{F}) \quad \text{instead of } \Delta z = -M(z)\hat{F},$$

where we have dropped the iteration index k for clarity. We define $J^{\oplus}$ implicitly in two steps. The fist step consists of computation of the primal-dual solution $\widetilde{z} = (\widetilde{x}, \widetilde{y}) \in \mathbb{R}^{n+m}$ of the QP

$$\underset{\widetilde{x} \in \mathbb{R}^n}{\text{minimize}} \quad \frac{1}{2}\widetilde{x}^{\mathrm{T}} J_1(z)\widetilde{x} + \left(\hat{F}_1 - J_1(z)x + J_2(z)^{\mathrm{T}} y\right)^{\mathrm{T}} \widetilde{x} \tag{5.32a}$$

$$\text{s.t.} \quad \left(J_2(z)\widetilde{x} + \left(\hat{F}_2 - J_2(z)x\right)\right)_i = 0, \qquad i \in \mathscr{E}, \tag{5.32b}$$

$$\left(J_2(z)\widetilde{x} + \left(\hat{F}_2 - J_2(z)x\right)\right)_i \geq 0, \qquad i \in \mathscr{I}, \tag{5.32c}$$

which is not formulated in the space of increments $\Delta z = (\Delta x, \Delta y) \in \mathbb{R}^{n+m}$ but rather in the space of variables $\widetilde{z} = z + \Delta z$. In the second step we reverse this transformation and obtain $\Delta z = \widetilde{z} - z$.

Lemma 5.36. *Assume that QP (5.32) at $z \in \mathbb{R}^N$ has a unique solution. If $\hat{z} = (\hat{x}, \hat{y}) \in \mathbb{R}^N$ satisfies $\hat{y}_i \geq -y_i$ for $i \in \mathscr{I}$ then*

$$J^{\oplus}(z, J(z)\hat{z}) = \hat{z}.$$

Proof. Let $\hat{z} = (\hat{x}, \hat{y}) \in \mathbb{R}^N$ be given and define

$$\hat{F} = -J(z)\hat{z} = \begin{pmatrix} -J_1(z)\hat{x} + J_2(z)\hat{y} \\ -J_2(z)\hat{x} \end{pmatrix}.$$

To prove the lemma we show that $J^{\oplus}(z, -\hat{F}) = \Delta z = \hat{z}$. With aforementioned choice of $\hat{F}$ QP (5.32) becomes

$$\underset{\widetilde{x} \in \mathbb{R}^n}{\text{minimize}} \quad \frac{1}{2}\widetilde{x}^{\mathrm{T}} J_1(z)\widetilde{x} - \left(J_1(z)(\hat{x}+x) - J_2(z)^{\mathrm{T}}(\hat{y}+y)\right)^{\mathrm{T}} \widetilde{x}$$

$$\text{s.t.} \quad J_2(z)_i (\widetilde{x} - \hat{x} - x) = 0, \qquad i \in \mathscr{E},$$

$$J_2(z)_i (\widetilde{x} - \hat{x} - x) \geq 0, \qquad i \in \mathscr{I}.$$

Its stationarity condition reads

$$J_1(z)(\widetilde{x} - \hat{x} - x) - J_2(z)(\widetilde{y} - \hat{y} - y) = 0.$$

We thus observe that $\widetilde{z} = \hat{z} + z$ is stationary and primal feasible. Dual feasibility holds due to $\widetilde{y}_i = \hat{y}_i + y_i \geq 0$ for $i \in \mathscr{I}$ by assumption. Complementarity is satisfied by virtue of $\widetilde{x} - \hat{x} - x = 0$. Thus $\Delta z = \widetilde{z} - z = \hat{z}$. $\square$

Lemma 5.36 reveals that under the stated assumptions $J^{\oplus}$ operates linear on the second argument like a generalized inverse of J.

Theorem 5.37. *Assume that $\alpha_{k-1} = 1$ and that the solutions of QP (5.32) at $(z^{k-1}, -F(z^{k-1}))$ and $(z^k, -F(z^k))$ share the same active set $\mathscr{A}$ and satisfy the SOSC and the SCC. Then there exists a matrix M^k and a neighborhood U of $F(z^k)$ such that*

$$-M^k \hat{F} = J^\oplus(z^k, -\hat{F}) \quad \text{for all } \hat{F} \in U.$$

Proof. We first notice that the solution $\tilde{z}^{k-1}$ of QP (5.32) at $(z^{k-1}, -F(z^{k-1}))$ satisfies

$$z^k = z^{k-1} + \Delta z^{k-1} = \tilde{z}^{k-1}.$$

Thus we have for all inactive inequality constraints that

$$y_i^k = \tilde{y}_i^{k-1} = 0 \quad \text{and} \quad y_i^{k+1} = \tilde{y}_i^k = 0 \quad \text{for } i \in \overline{m} \setminus \mathscr{A}$$

by virtue of complementarity. It follows that $\Delta y_i^k = 0, i \in \overline{m} \setminus \mathscr{A}$ and thus we can set all rows $n + i, i \in \overline{m} \setminus \mathscr{A}$ of M^k to zero. Due to invariance of the active set $\mathscr{A}$ we obtain for the remaining variables the linear system

$$\begin{pmatrix} J_1(z^k) & -J_2(z^k)_{\mathscr{A}}^{\mathrm{T}} \\ J_2(z^k)_{\mathscr{A}} & 0 \end{pmatrix} \begin{pmatrix} x^{k+1} \\ y_{\mathscr{A}}^{k+1} \end{pmatrix} + \begin{pmatrix} F_1(z^k) \\ F_2(z^k)_{\mathscr{A}} \end{pmatrix} = 0, \qquad (5.33)$$

whose solution depends linearly on $F(z^k)$ and defines the submatrix of M^k corresponding to primal and active dual variables. We further notice that z^{k+1} does not depend on $F_2(z^k)_{\overline{m} \setminus \mathscr{A}}$. Thus we can set all remaining columns $n + i, i \in \overline{m} \setminus \mathscr{A}$ of M^k to zero. This fully defines the matrix M^k.

Because the SOSC and the SCC hold, the active set $\mathscr{A}$ is stable under perturbations (see, e.g., Robinson [132]) which yields the existence of a neighborhood U of $F(z^k)$ such that

$$-M^k \hat{F} = J^\oplus(z^k, -\hat{F}) \quad \text{for all } \hat{F} \in U.$$

This completes the proof. □

The proof of Theorem 5.37 explicitly constructs a matrix $M(z^k)$ as the linearization of $J^\oplus(z^k, .)$ around $-F(z^k)$ which exists under the stated assumptions. Thus we can invoke the Local Contraction Theorem 5.5 if the solution z^* satisfies the SOSC and the SCC.

Remark 5.38. In the case of varying active sets between two consecutive QPs the action of $J^\oplus(z^k, -F(z^k))$ can be interpreted as an affine linear function consisting of an offset for Δz plus a linear term $-M^k F(z^k)$, where M can be constructed like in Theorem 5.37 with a small enough step size $\alpha_k > 0$ such that $F(z^{k+1}) \in U$. From a geometrical point of view the overall iteration takes place on nonlinear segments given by the QP active sets with jumps between these segments. The assumption that the reduced Jacobian given in equation (5.33) is invertible on each segment is now as unrestrictive as the assumption of invertibility of $J(z^k)$ for the root finding problem $F(z) = 0$.

Remark 5.39. Algorithmically, the evaluation of M^k is performed in the following way: If M^k is evaluated for the first time, a full QP (5.32) is solved. For all further evaluations the active (or working) set is kept fixed and a purely equality constrained QP is solved.

Remark 5.40. We are not aware of results how the jumps due to $J^\oplus$ can be analyzed within the non-local theory developed in Section 5.6 for globalization based on an NMT. We have not yet attempted an approach to fill this gap yet, either. However, the numerical results that we present in Part III are encouraging to undertake such a probably difficult endeavor.

The following theorem ensures that limit points of the SQP iteration with $J^\oplus$ are indeed KKT points or even local solutions if SOSC holds on the QP level.

Theorem 5.41. *If the SQP method with $J^\oplus$ converges to z^* then z^* is a KKT point of NLP (4.1). Furthermore, the conditions SOSC and SCC transfer from QP (5.32) at z^* to NLP (4.1) (at z^*).*

Proof. If the SQP method converges it must hold that

$$0 = -M(z^*)F(z^*) = J^\oplus(z^*, F(z^*)),$$

i.e., $\widetilde{z} = z^*$ is a solution of

$$\begin{aligned}
\underset{\widetilde{x}\in\mathbb{R}^n}{\text{minimize}} \quad & \frac{1}{2}\widetilde{x}^\mathrm{T} J_1(z^*)\widetilde{x} + \left(F_1(z^*) - J_1(z^*)x^* + J_2(z^*)^\mathrm{T} y^*\right)^\mathrm{T}\widetilde{x} \\
\text{s.t.} \quad & \left(J_2(z^*)\widetilde{x} + F_2(z^*) - J_2(z^*)x^*\right)_i = 0, & i \in \mathscr{E}, \\
& \left(J_2(z^*)\widetilde{x} + F_2(z^*) - J_2(z^*)x^*\right)_i \geq 0, & i \in \mathscr{I}.
\end{aligned}$$

We immediately observe primal feasibility for $F_2(z^*) = g(z^*)$. From QP stationarity we obtain NLP stationarity by virtue of

$$0 = J_1(z^*)\widetilde{x} + F_1(z^*) - J_1(z^*)x^* + J_2(z^*)^\mathrm{T} y^* - J_2(z^*)^\mathrm{T}\widetilde{y} = F_1(z^*) = \nabla_x\mathscr{L}(z^*).$$

Dual feasibility and complementarity for the NLP as well as SOSC and SCC follow directly from the QP. $\square$

5.7.2 Inexact SQP

The goal of this section is to present how the application of an approximation of $J^{\oplus}$ within a LISA-Newton method (see Section 5.6) can be evaluated. Let us assume that we have an approximation of the Jacobian matrix (e.g., via a Newton-Picard approximation described in Chapter 6) given by

$$J(z^k) = \begin{pmatrix} \nabla^2_{xx}\mathscr{L}(z^k) & -\nabla g(x^k) \\ \nabla g(x^k)^{\mathrm{T}} & 0 \end{pmatrix} \approx \begin{pmatrix} B^k & -(C^k)^{\mathrm{T}} \\ C^k & 0 \end{pmatrix} =: \hat{J}^k.$$

We perform the construction of a preconditioner $\hat{M}(z^k)$ for LISA based on $\hat{J}^k$ now analogously to the construction of $J^{\oplus}(z^k,.)$ from $J(z^k)$. The key point is that the transformation now requires the sum of the current Newton and the current LISA iterate $z^k + \delta z_l^k$. Dropping the index k we solve the QP

$$\underset{\tilde{x}\in\mathbb{R}^n}{\text{minimize}} \quad \frac{1}{2}\tilde{x}^{\mathrm{T}}B\tilde{x} + \left(\hat{F}_1 - C(x+\delta x_l) + C^{\mathrm{T}}(y+\delta y_l)\right)^{\mathrm{T}}\tilde{x} \tag{5.34a}$$

$$\text{s.t.} \quad \left(C\tilde{x} + \left(\hat{F}_2 - C(x+\delta x_l)\right)\right)_i = 0, \qquad\qquad i \in \mathscr{E}, \tag{5.34b}$$

$$\left(C\tilde{x} + \left(\hat{F}_2 - C(x+\delta x_l)\right)\right)_i \geq 0, \qquad\qquad i \in \mathscr{I}, \tag{5.34c}$$

and reverse the transformation afterwards via $\Delta z^k = \tilde{z}^k - z^k - \delta z_l^k$.

Remark 5.39 about the evaluation of M^k is also valid in the context of inexact SQP for $\hat{M}^k$.

Algorithm 1: Global inexact Newton method with κ and ω monitoring (GINKO)

evaluate $F_0 = F(z_0)$, set $\delta z_0^0 = 0, k = 0, l = 0$

k   **if** $k \geq k_{\max}$ **then** *Error: Maximum outer Newton iterations reached*

l   **if** $l \geq l_{\max}$ **then** *Error: Maximum iterations for κ improvement reached*

  set $j = 0, \kappa = 0, \delta z^k =$ not found

j   **if** $\delta z^k \neq$ not found **then**

    | **if** $\alpha_k < \alpha_{\min}$ **then** *Error: Minimum step size reached*
    | set $z^{k+1} = z^k + \alpha_k \delta z^k, \delta z_0^k = (1 - \alpha_k) \delta z^k$
    | evaluate $F_{k+1} = F(z^{k+1})$

  set $i = 0$

i   **if** $i \geq i_{\max}$ **then** *Error: Maximum inner iterations reached*

  if $\delta z^k =$ not found **then** compute residual $r_i^k = -F_k - J(z_k)\delta z_i^k$

  else compute residual $r_i^k = -F_{k+1} - r^k - J(z_k)\delta z_i^k$

  refine increment iterate $\delta z_{i+1}^k = \delta z_i^k + \hat{M}(z_k)r_i^k$

  if $i < 1$ **then** set $i = i + 1$ and **goto** i

  estimate contraction $[\hat{\kappa}] \approx \hat{\kappa}$

  if $i > i_{pre}$ *and* $[\hat{\kappa}] > \kappa_{\max}$ **then** ameliorate $\hat{M}(z_k)$, set $l = l + 1$, and **goto** l

  if $\delta z^k =$ not found **then**

    | estimate error $\delta_k^{i+1} = [\hat{\kappa}] \left\| \delta z_{i+1}^k - \delta z_i^k \right\| / \left((1 - [\hat{\kappa}]) \left\| \delta z_{i+1}^k \right\| \right)$
    | **if** $[(\alpha < 1) \wedge (\delta_k^{i+1} > \rho/(2(1+\rho)))] \vee [(\alpha = 1) \wedge (\delta_k^{i+1} > (\rho/2)[h_k^\delta])]$ **then**
    | | set $i = i + 1$ and **goto** i
    | set $\delta z^k = \delta z_{i+1}^k$ and save $i, \delta z_i^k, \delta z_{i+1}^k$
    | **if** $\left\| \delta z^k \right\| <$ TOL **then** terminate with solution $z^{k+1} = z^k + \delta z^k$
    | compute residual $r^k = -F_k - J(z_k)\delta z^k$
    | **if** $k > 0$ **then**
    | | adapt a-priori Kantorovich estimate $[h_k^\delta] = (\left\| \delta z^k \right\| / \left\| \delta z^{k-1} \right\|)[h_{k-1}^\delta]_*$
    | | adapt step size $\alpha_k = \max(\min(1, 1/((1+\rho)[h_k^\delta])), \alpha_{\maxred}\alpha_{k-1})$
    | set $\delta_k^* = \left\| \delta z_{i+1}^k - \delta z_i^k \right\| / \left\| \delta z_{i+1}^k \right\|, \delta z_0^k = \delta z^k, j = 0$ and **goto** j

  else

    | recheck accuracy of δz^k: $\delta_k = [\hat{\kappa}]\delta_k^*/(1 - [\hat{\kappa}])$
    | **if** $[(\alpha < 1) \wedge (\delta_k > \rho/(2(1+\rho)))] \vee [(\alpha = 1) \wedge (\delta_k > (\rho/2)[h_k^\delta])]$ **then**
    | | restore $i, \delta z_i^k, \delta z_{i+1}^k$, set $\delta z^k =$ not found, $i = i + 1$, and **goto** i
    | compute $\bar{\rho}_{i+1} = [\hat{\kappa}] \left\| \delta z_{i+1}^k - \delta z_i^k \right\| / \left((1 - [\hat{\kappa}]) \left\| \delta z_{i+1}^k - (1 - \alpha_k)\delta z^k \right\| \right)$
    | **if** $\bar{\rho}_{i+1} > \bar{\rho}_{\max}$ **then** set $i = i + 1$ and **goto** i
    | a-posteriori estimate $[h_k^\delta]_* = 2(1 - \bar{\rho}_{i+1}) \left\| \delta z_{i+1}^{k+1} - \delta z_0^{k+1} \right\| / (\alpha_k^2 \left\| \delta z^k \right\|)$
    | compute monitor $\Theta_k = \left\| \delta z^{k+1} \right\| / \left\| \delta z^k \right\|$
    | **if** $\Theta_k \geq 1 - (\rho - \tilde{\rho})/(1 + \rho)$ **then**
    | | adapt step $\alpha_k = \max(\min(1, 1/((1+\rho)[h_k^\delta]_*)), \alpha_{\maxred}\alpha_{k-1})$ and **goto** j
    | set $\delta z_0^{k+1} = \delta z_{i+1}^k, k = k + 1, l = 0$, and **goto** k

---

6 Newton-Picard preconditioners

For completeness we give the following excerpt from the preprint Potschka et al. [131] here with adaptions in the variable names to fit the presentation in this thesis.

We present preconditioners for the iterative solution of symmetric indefinite linear systems

$$\hat{J}z = \begin{pmatrix} \hat{J}_1 & \hat{J}_2^{\mathrm{T}} \\ \hat{J}_2 & 0 \end{pmatrix} \begin{pmatrix} x \\ y \end{pmatrix} = - \begin{pmatrix} \hat{F}_1 \\ \hat{F}_2 \end{pmatrix} =: -\hat{F},$$

with $\hat{J} \in \mathbb{R}^{(n+m)\times(n+m)}$, $z, \hat{F} \in \mathbb{R}^{n+m}$ derived from equation (5.33) within the framework of an SQP method (see Chapter 5). Note that we have swapped the sign of y to achieve symmtery of $\hat{J}$. It is well known (see, e.g., Nocedal and Wright [121]) that $\hat{J}$ is invertible if $\hat{J}_2$ has full rank and $\hat{J}_1$ is positive definite on the nullspace of $\hat{J}_1$. For weaker sufficient conditions for invertibility of $\hat{J}$ and a survey of solution techniques we refer the reader to Benzi et al. [16].

We base our investigations on the following linear-quadratic model problem: Let $\Omega \subset \mathbb{R}^d$ be a bounded open domain with Lipschitz boundary $\partial\Omega$ and let $\Sigma := (0,1) \times \partial\Omega$. We seek controls $q \in L^2(\Sigma)$ and corresponding states $u \in W(0,1)$ which solve the time-periodic PDE OCP

$$\underset{q\in L^2(\Sigma), u\in W(0,1)}{\text{minimize}} \quad \frac{1}{2} \int_\Omega (u(1;.) - \hat{u})^2 + \frac{\gamma}{2} \iint_\Sigma q^2 \tag{6.1a}$$

$$\text{s.t.} \quad \partial_t u = D\Delta u \quad \text{in } (0,1)\times\Omega, \tag{6.1b}$$

$$\partial_\nu u + \alpha u = \beta q \quad \text{in } (0,1)\times\partial\Omega, \tag{6.1c}$$

$$u(0;.) = u(1;.) \quad \text{in } \Omega, \tag{6.1d}$$

with $\hat{u} \in L^2(\Omega)$, $\alpha, \beta \in L^\infty(\partial\Omega)$ non-negative a.e., $\|\alpha\|_{L^\infty(\partial\Omega)} > 0$, and $D, \gamma > 0$. This problem is an extension of the parabolic optimal control problem presented, e.g., in the textbook of Tröltzsch [152].

Our focus here lies on splitting approaches

$$\hat{J} = \tilde{J} - \Delta J$$

with $\tilde{J}, \Delta J \in \mathbb{R}^{(n_1+n_2)\times(n_1+n_2)}$ and $\tilde{J}$ invertible. We employ these splittings in a LISA (see Chapter 5) which has the form

$$z^{k+1} = z^k - \tilde{J}^{-1}(\hat{J}z^k + \hat{F}) = \tilde{J}^{-1}\Delta J z^k - \tilde{J}^{-1}\hat{F}. \tag{6.2}$$

As a guiding principle, the iterations should not be forced to lie on the subset of *feasible* (possibly *non-optimal*) points, which satisfy $\hat{J}_2 x^k = -\hat{F}_2$ for all k, i.e., the PDE constraints are allowed to be violated in iterates away from the optimal solution. Instead, feasibility and optimality are supposed to hold only at the solution. The presence or absence of this property defines the terms *sequential/segregated* and *simultaneous/all-at-once/coupled* method, whereby a method with only feasible iterates is called sequential or segregated. The preconditioners we present work on formulations of the problem which lead to simultaneous iterations. From a computational point of view, simultaneous methods are more attractive because it is not necessary to find an exact solution of $\hat{J}_2 x^k = -\hat{F}_2$ in every iteration.

This chapter is organized as follows: In Section 6.1 we give a short review of the Newton-Picard related literature. We discuss the discretization of problem (6.1) afterwards in Section 6.2. In Section 6.3 we present the Newton-Picard preconditioners in the framework of LISA (see Chapter 5). For the discretized problem we discuss the cases of classical Newton-Picard projective approximation and of a coarse-grid approach for the constraint Jacobians. The importance of the choice of the scalar product for the projection is highlighted. We establish a mesh-independent convergence result for LISA based on classical Newton-Picard splitting. In this section we also outline the fast solution of the subproblems, present pseudocode, and analyze the computational complexity. Moreover we discuss extensions to nonlinear problems and the Multiple Shooting case in Section 6.4.

In Chapter 12 of this thesis we present numerical results for different sets of problem and discretization parameters for the Newton-Picard preconditioners. In addition, we compare the indefinite Newton-Picard preconditioners with a symmetric positive definite Schur complement preconditioner in a Krylov method setting.

6.1 The Newton-Picard method for finding periodic steady states

In the context of bifurcation analysis of large nonlinear systems Jarausch and Mackens [90] have developed the so-called Condensed Newton with Supported Picard approach to solve fixed point equations which have a few unstable or slowly converging modes. Their presentation is restricted to systems with symmetric Jacobian. Shroff and Keller [147] extended the approach to the unsymmetric case with the Recursive Projection Method by using more sophisticated numerical methods for the identification of the slow eigenspace. There are two articles in volume 19(4) of the SIAM Journal on Scientific Computing which are both based on [90, 147]: Lust et al. [110] successfully applied the Newton-Picard method for computation and bifurcation analysis of time-periodic solutions of PDEs and Burrage et al. [32] develop the notion of deflation preconditioners. To our knowledge the first paper on deflation techniques is by Nicolaides [119] who explicitly introduces deflation as a modification to the conjugate gradient method and not as a preconditioner in order to improve convergence. Fast two-grid and multigrid approaches for the determination of time-periodic solutions for parabolic PDEs on the basis of a shooting approach are due to Hackbusch [77].

6.2 Discretization of the model problem

A full space-time discretization of problem (6.1) would lead to prohibitively large memory requirements for $d = 3$. Thus, we employ a shooting approach which reduces the degrees of freedom for the state variables to only the initial value. Let us recapitulate the discretization steps outlined in Chapter 3 and apply them to the model problem (6.1). We discretize the controls in space with n_q form functions $\tilde{\psi}_l$ whose amplitude can be controlled in time, i.e.,

$$q(t,x) = \sum_{l=1}^{n_q} q_l(t)\tilde{\psi}_l(x), \quad q_l \in L^2(0,1), \tilde{\psi}_l \in L^2(\partial\Omega).$$

In weak variational form a solution $u \in W(0,1)$ of PDE (6.1b) satisfies for all $\varphi \in H^1(\Omega)$ and almost all $t \in [0,1]$ the equation

$$\int_\Omega u_t(t)\varphi = -D\int_\Omega \nabla u(t)^\mathrm{T}\nabla\varphi + D\int_{\partial\Omega} \partial_\nu u(t)\varphi \tag{6.3a}$$

$$= -D\int_\Omega \nabla u(t)^\mathrm{T}\nabla\varphi - D\int_{\partial\Omega} \alpha u(t)\varphi + D\int_{\partial\Omega} \beta q(t)\varphi. \tag{6.3b}$$

We continue with discretizing the state u in space using a Galerkin approach. Let $\varphi_i \in H^1(\Omega)$, $i = 1,\ldots,n_u$, denote linearly independent functions, e.g., FEM hat functions on a mesh with n_u vertices, and define the matrices $S,Q,M \in \mathbb{R}^{n_u \times n_u}, U \in \mathbb{R}^{n_u \times n_q m}$ and the vector $\hat{u} \in \mathbb{R}^{n_u}$ according to

$$S_{ij} = D\int_\Omega \nabla\varphi_i^\mathrm{T}\nabla\varphi_j, \quad Q_{ij} = D\int_{\partial\Omega} \alpha\varphi_i\varphi_j, \quad U_{il} = D\int_{\partial\Omega} \beta\varphi_i\tilde{\psi}_l,$$

$$M_{ij} = \int_\Omega \varphi_i\varphi_j, \quad \hat{u}_i = \int_\Omega \hat{u}\varphi_i.$$

It is well known that the mass matrix M is symmetric positive definite. We can now discretize equation (6.3) with MOL: The matrix of the discretized spatial differential operator is $L = -S - Q$ which leads to the Ordinary Differential Equation (ODE)

$$M\dot{u}(t) = Lu(t) + U(q_1(t) \cdots q_{n_q}(t))^\mathrm{T}, \tag{6.4}$$

where $u(t) = \sum_{i=1}^{n_u} u_i(t)\varphi_i$. Then, we discretize each $q_l(t)$ using piecewise constant functions on m intervals. We uniquely decompose $i = i_t m + i_q$ with $i_t = \lfloor (i-1)/m \rfloor$ and $i_q = i - i_t m$. Thus, $0 \le i_t < m, 1 \le i_q \le n_q$ and we can define

$$\psi_i(t,x) = \chi_{[i_t/m,(i_t+1)/m]}(t)\tilde{\psi}_{i_q}(x) \in L^2(\Sigma)$$

as a basis for the discrete control space. Here χ denotes the characteristic function of the subscript interval. We can then define the symmetric positive definite control mass matrix $N \in \mathbb{R}^{n_q m \times n_q m}$ according to

$$N_{ij} = \iint_{\Sigma} \psi_i \psi_j.$$

Moreover we denote the discretized controls by $q \in \mathbb{R}^{n_q m}$.

It is well known that the end state $u(1)$ depends linearly on $u(0)$ and q due to linearity of ODE (6.4). Thus, there exist unique matrices $G_u \in \mathbb{R}^{n_u \times n_u}$ and $G_q \in \mathbb{R}^{n_u \times n_q m}$ such that

$$u(1) = G_u u(0) + G_q q.$$

Now we construct formulas for G_u and G_q. We first consider solutions of ODE (6.4) for initial value $u(0) = u_0$ and controls $\tilde{q} \in \mathbb{R}^{n_q}$ which are constant in time. We can easily verify that the solution is given by the expression

$$u(t) = \exp(tM^{-1}L)u_0 + \left(\exp(tM^{-1}L) - \mathbb{I}_{n_u}\right)L^{-1}U\tilde{q}, \qquad (6.5)$$

where $\mathbb{I}_{n_u}$ denotes the n_u-by-n_u identity matrix. If we consider the special case $\tilde{q} = 0$ we immediately observe that matrix G_u is given by the matrix exponential

$$G_u = \exp(M^{-1}L). \qquad (6.6)$$

Because ODE (6.4) is autonomous the matrix G_q can be composed piece by piece on the control time grid based on the matrices $\partial G_u := \exp((1/m)M^{-1}L)$ and $\partial G_q := (\partial G_u - \mathbb{I}_{n_u})L^{-1}U$ for a single interval. We obtain

$$G_q = \left(\partial G_u^{m-1} \partial G_q \quad \cdots \quad \partial G_u^1 \partial G_q \quad \partial G_u^0 \partial G_q\right). \qquad (6.7)$$

We now investigate spectral properties of G_u. We start by showing that the unsymmetric matrix $M^{-1}L$ has a basis of M-orthonormal real eigenvectors and only real eigenvalues.

Lemma 6.1. *There exists an invertible matrix $Z \in \mathbb{R}^{n_u \times n_u}$ and a diagonal matrix $\tilde{E} \in \mathbb{R}^{n_u \times n_u}$ such that*

$$Z^{\mathsf{T}}MZ = \mathbb{I}_{n_u} \quad and \quad M^{-1}LZ = Z\tilde{E}.$$

Proof. The matrix $L = -S - Q$ is symmetric as a sum of symmetric Galerkin matrices. We decompose

$$M = R_M^{\mathrm{T}} R_M$$

with invertible $R_M \in \mathbb{R}^{n_u \times n_u}$, e.g., by Cholesky decomposition, and use matrix R_M for the equivalence transformation

$$R_M (M^{-1} L) R_M^{-1} = R_M^{-\mathrm{T}} L R_M^{-1}$$

of $M^{-1} L$ to a symmetric matrix. Thus, there exists an invertible matrix $\tilde{Z} \in \mathbb{R}^{n_u \times n_u}$ of eigenvectors of $R_M^{-\mathrm{T}} L R_M^{-1}$ satisfying $\tilde{Z}^{\mathrm{T}} \tilde{Z} = \mathbb{I}_{n_u}$ and a diagonal real matrix of eigenvalues $\tilde{E} \in \mathbb{R}^{n_u \times n_u}$ such that

$$R_M^{-\mathrm{T}} L R_M^{-1} \tilde{Z} = \tilde{Z} \tilde{E} \quad \text{(or, equivalently, } R_M^{-1} R_M^{-\mathrm{T}} L R_M^{-1} \tilde{Z} = R_M^{-1} \tilde{Z} \tilde{E}).$$

We define $Z := R_M^{-1} \tilde{Z}$ and immediately obtain the assertions. $\square$

Now we prove a negative upper bound on the eigenvalues of $M^{-1} L$.

Lemma 6.2. *There exists a grid-independent scalar $\bar{\mu} < 0$ such that all eigenvalues $\tilde{\mu}$ of $M^{-1} L$ satisfy $\tilde{\mu} \leq \bar{\mu}$.*

Proof. Let $(v, \tilde{\mu}) \in \mathbb{R}^{n_u} \times \mathbb{R}, v \neq 0$, be an eigenpair of $M^{-1} L$

$$M^{-1} L v = \tilde{\mu} v,$$

and define $v = \sum_{i=1}^{n_u} v_i \varphi_i \in H^1(\Omega)$. We now follow a step in a proof of Tröltzsch [152, Satz 2.6]: By the assumption of $\|\alpha\|_{L^\infty(\partial\Omega)} > 0$ there exists a measurable subset $\Gamma \subset \partial\Omega$ with positive measure and a scalar $\delta > 0$ with $\alpha \geq \delta$ a.e. on Γ. We obtain

$$\tilde{\mu} \|v\|_{L^2(\Omega)}^2 = \tilde{\mu} v^{\mathrm{T}} M v = v^{\mathrm{T}} L v$$

$$= -D \left(\int_\Omega \nabla v^{\mathrm{T}} \nabla v + \int_{\partial\Omega} \alpha v^2 \right) \leq -D \left(\int_\Omega \|\nabla v\|_2^2 + \delta \int_\Gamma v^2 \right).$$

Then we apply the generalized Friedrichs inequality [152, Lemma 2.5] which yields a Γ-dependent constant $c(\Gamma) > 0$ such that

$$\tilde{\mu} \|v\|_{L^2(\Omega)}^2 \leq -D \left(\int_\Omega \|\nabla v\|_2^2 + \delta \int_\Gamma v^2 \right) \leq \frac{D \min(1, \delta)}{c(\Gamma)} \left(-\|v\|_{H^1(\Omega)}^2 \right).$$

With $-\|v\|_{H^1(\Omega)}^2 \leq -\|v\|_{L^2(\Omega)}^2$ we finally obtain the assertion for the choice $\bar{\mu} := -D \min(1, \delta)/c(\Gamma) < 0$. $\square$

Lemma 6.3. *Let $\mu \in \mathbb{C}$ be an eigenvalue of G_u. Then μ is real and there exists a grid-independent scalar $\bar{\mu} < 1$ such that $0 < \mu \leq \bar{\mu}$.*

Proof. The matrix G_u has the same eigenvectors as the matrix $M^{-1}L$. Thus, the assertion is a direct consequence of equation (6.6) and Lemma 6.2 with $\bar{\mu} = \exp(\bar{\tilde{\mu}}) \in (0, 1)$. $\square$

We now formulate the finite dimensional linear-quadratic optimization problem

$$\underset{u_0 \in \mathbb{R}^{n_u}, q \in \mathbb{R}^{n_q m}}{\text{minimize}} \frac{1}{2} u_0^T M u_0 - \hat{u}^T u_0 + \gamma q^T N q \tag{6.8a}$$

$$\text{s.t.} \qquad M(G_u - \mathbb{I}_{n_u}) u_0 + M G_q q = 0. \tag{6.8b}$$

Lemma 6.4. *Problem (6.8) has a unique solution.*

Proof. Due to convexity of problem (6.8), necessary optimality conditions are also sufficient, i.e., if there exists a multiplier vector $\lambda \in \mathbb{R}^{n_u}$ such that for $u_0 \in \mathbb{R}^{n_u}, q \in \mathbb{R}^{n_q m}$ it holds that

$$\begin{pmatrix} M & 0 & (G_u^T - \mathbb{I}_{n_u})M \\ 0 & \gamma N & G_q^T M \\ M(G_u - \mathbb{I}_{n_u}) & M G_q & 0 \end{pmatrix} \begin{pmatrix} u_0 \\ q \\ \lambda \end{pmatrix} = \begin{pmatrix} \hat{u} \\ 0 \\ 0 \end{pmatrix} \tag{6.9}$$

then (u_0, q, λ) is a primal-dual optimal solution and, conversely, all optimal solutions must satisfy condition (6.9). The constraint Jacobian

$$M \begin{pmatrix} G_u - \mathbb{I}_{n_u} & G_q \end{pmatrix}$$

has full rank due to $G_u - \mathbb{I}_{n_u}$ being invertible by virtue of Lemma 6.3. The Hessian blocks M and γN are positive definite. Thus, the symmetric indefinite linear system (6.9) is non-singular and has a unique solution. $\square$

6.3 Newton-Picard for optimal control problems

In this section we investigate how the Newton-Picard method for the forward problem (i.e., solving for a periodic state for given controls) can be exploited in a simultaneous optimization approach.

6.3.1 General considerations

For large values of n_u it is prohibitively expensive to explicitly form the matrix in equation (6.9) because the matrix G_u is a large, dense n_u-by-n_u matrix. Thus, we cannot rely on direct linear algebra for the solution of equation (6.9). However, we observe that matrix-vector products are relatively economical to evaluate: The cost of an evaluation of $G_u v$ is the cost of a numerical integration of ODE (6.4) with initial value v and controls $q = 0$. The evaluation of $G_u^T v$ can be computed using the identities

$$MG_u = M\exp(M^{-1}L) = \exp(LM^{-1})M = G_u^T M, \quad G_u^T v = MG_u M^{-1} v. \quad (6.10)$$

Matrix vector products with G_q and G_q^T can then be evaluated based on equation (6.7).

The main difficulty here are the large and dense G_u blocks and thus approaches based on the paper of Bramble and Pasciak [30] and also constraint preconditioners (e.g., Gould et al. [67]), which do not approximate the blocks containing G_u but only the M and γN blocks, do not attack the main difficulty of the problem and will thus be not considered further in this thesis.

6.3.2 Simultaneous Newton-Picard iteration

LISA for the linear system (6.9) yields a simultaneous optimization method because the iterations will in general not satisfy the periodicity constraint before convergence. The type of preconditioners we study here is of the following form: Let $\tilde{G}_u$ denote an approximation of G_u and regard the exact and approximated matrices

$$\hat{J} := \begin{pmatrix} M & 0 & (G_u^T - \mathbb{I}_{n_u})M \\ 0 & \gamma N & G_q^T M \\ M(G_u - \mathbb{I}_{n_u}) & MG_q & 0 \end{pmatrix},$$

$$\tilde{J} := \begin{pmatrix} M & 0 & (\tilde{G}_u^T - \mathbb{I}_{n_u})M \\ 0 & \gamma N & G_q^T M \\ M(\tilde{G}_u - \mathbb{I}_{n_u}) & MG_q & 0 \end{pmatrix}.$$

We investigate two choices for $\tilde{G}_u$: The first is based on the classical Newton-Picard projective approximation [110] for the forward problem, the second is based on a two-grid idea.

6.3.2.1 Classical Newton-Picard projective approximation

The principle of the Newton-Picard approximation is based on observations about the spectrum of the monodromy matrix G_u (see Figure 12.1 in Chapter 12 on page 164). The eigenvalues μ_i cluster around zero and there are only few eigenvalues that are close to the unit circle. The cluster is a direct consequence of the dissipativity of the underlying heat equation, i.e., high-frequency components in space get damped out rapidly. Thus, the zero matrix is a good approximation of G_u in directions of eigenvectors corresponding to small eigenvalues. The rationale behind a Newton-Picard approximation consists of approximating G_u exactly on the low-dimensional space of eigenvectors corresponding to large eigenvalues. To this end, let the columns of the orthonormal matrix $V \in \mathbb{R}^{n_u \times p}$ be the p eigenvectors of G_u with largest eigenvalues μ_i such that

$$G_u V = VE, \quad E \in \mathbb{R}^{p \times p} \text{ diagonal.}$$

Now, we approximate the matrix G_u with

$$\tilde{G}_u = G_u \Pi,$$

where Π is a projector onto the dominant subspace of G_u. Lust et al. [110] proposed to use

$$\Pi = VV^{\mathrm{T}}, \tag{6.11}$$

which is an orthogonal projector in the Euclidean sense. This works well for the solution of the pure forward problem but in a simultaneous optimization approach, this choice may lead to undesirable loss of contraction, as shown in Chapter 12. We propose to use a projector that instead takes the scalar product of the infinite dimensional space into account. The projector maps a vector $w \in \mathbb{R}^{n_u}$ to the closest point $Vv, v \in \mathbb{R}^p$, of the dominant subspace in an L^2 sense, by solving the minimization problem

$$\underset{v \in \mathbb{R}^p}{\text{minimize}} \; \frac{1}{2} \|w - v\|^2_{L^2(\Omega)} = \frac{1}{2} v^{\mathrm{T}} V^{\mathrm{T}} MVv - v^{\mathrm{T}} V^{\mathrm{T}} Mw + \frac{1}{2} w^{\mathrm{T}} Mw,$$

where $w = \sum_{i=1}^{n_u} w_i \varphi_i$ and $v = \sum_{i=1}^{n_u} (Vv)_i \varphi_i$. The projector is therefore given by

$$\Pi = VM_p^{-1}V^{\mathrm{T}}M, \quad \text{where } M_p = V^{\mathrm{T}}MV \in \mathbb{R}^{p \times p}. \tag{6.12}$$

Thus, we approximate G_u with

$$\tilde{G}_u = VEM_p^{-1}V^{\mathrm{T}}M.$$

To compute the inverse of $\tilde{G}_u - \mathbb{I}_{n_u}$ we have the following lemma which we invoke with $P = V$ and $R = M_p^{-1}V^{\mathrm{T}}M$:

Lemma 6.5. *Let* $\tilde{G}_u \in \mathbb{R}^{n_u \times n_u}, P \in \mathbb{R}^{n_u \times p}, R \in \mathbb{R}^{p \times n_u}$, *and* $E \in \mathbb{R}^{p \times p}$ *satisfy*

$$\tilde{G}_u = PER \quad and \quad RP = \mathbb{I}_p.$$

If $E - \mathbb{I}_p$ *is invertible then the inverse of* $\tilde{G}_u - \mathbb{I}_{n_u}$ *is given by*

$$(\tilde{G}_u - \mathbb{I}_{n_u})^{-1} = PXR - \mathbb{I}_{n_u}, \quad where \ X = (E - \mathbb{I}_p)^{-1} + \mathbb{I}_p.$$

Proof. Based on the Sherman-Morrison-Woodbury formula (see, e.g., Nocedal and Wright [121]) we obtain

$$(\tilde{G}_u - \mathbb{I}_{n_u})^{-1} = (-\mathbb{I}_{n_u} + PER)^{-1} = -\mathbb{I}_{n_u} - P(\mathbb{I}_p - ERP)^{-1}ER$$
$$= P(E - \mathbb{I}_p)^{-1}ER - \mathbb{I}_{n_u}.$$

The result follows from the identity $(E - \mathbb{I}_p)^{-1}(\mathbb{I}_p + E - \mathbb{I}_p) = (E - \mathbb{I}_p)^{-1} + \mathbb{I}_p = X$.
□

Computation of the inverse thus only needs the inversion of the small p-by-p matrices $E - \mathbb{I}_p$ and M_p. For inversion of $\tilde{G}_u^{\mathrm{T}} - \mathbb{I}_{n_u}$ we obtain similar to equation (6.10)

$$M\tilde{G}_u M^{-1} = MG_u V M_p^{-1} V^{\mathrm{T}} = G_u^{\mathrm{T}} M V M_p^{-1} V^{\mathrm{T}} = (G_u \Pi)^{\mathrm{T}} = \tilde{G}_u^{\mathrm{T}}$$

and consequently

$$(\tilde{G}_u^{\mathrm{T}} - \mathbb{I}_{n_u})^{-1} = \left(M(\tilde{G}_u - \mathbb{I}_{n_u})M^{-1} \right)^{-1} = M(\tilde{G}_u - \mathbb{I}_{n_u})^{-1}M^{-1}. \tag{6.13}$$

A dominant subspace basis V for the p-dimensional dominant eigenspace of $M^{-1}L$ and thus G_u can, e.g., be computed via an Implicitly Restarted Arnoldi Method (IRAM), see Lehoucq and Sorensen [103], for the (generalized) eigenvalue problem

$$M^{-1}LV - V\tilde{E} = 0 \quad \Leftrightarrow \quad LV - MV\tilde{E} = 0.$$

On the basis of equation (6.6) we obtain $E := \exp(\tilde{E})$.

6.3.2.2 Two-grid Newton-Picard

This variant is based on the observation that for the heat equation the slowly decaying modes are the low-frequency modes and the fast decaying modes are the high-frequency modes. Low-frequency modes can be approximated well on coarse grids. Thus we propose a method with two grids in which $\tilde{G}_u$ is calculated only on a coarse grid (cf. Hackbusch [77]), while the remaining computations are performed on the fine grid. Let P and R denote the prolongation and restriction matrices between the two grids and let superscripts c and f denote coarse and fine grid, respectively. Then, G_u^f is approximated by

$$\tilde{G}_u^f = PER, \quad \text{with } E := G_u^c,$$

i.e., we first project from the fine grid to the coarse grid, evaluate the exact G_u^c on the coarse grid, and prolongate the result back to the fine grid. Note that in contrast to classical Newton-Picard, E is now not a diagonal matrix.

We use conforming grids, i.e., the Finite Element basis on the coarse grid can be represented exactly in the basis on the fine grid. Thus, the prolongation P can be obtained by interpolation. Let $u^f \in \mathbb{R}^{n_u^f}, u^c \in \mathbb{R}^{n_u^c}$ and define

$$u^f = \sum_{i=1}^{n_u^f} u_i^f \varphi_i^f \in H^1(\Omega), \quad u^c = \sum_{i=1}^{n_u^c} u_i^c \varphi_i^c \in H^1(\Omega).$$

We define the restriction R in an L^2 sense, such that given u^f on the fine grid we look for the projector $R : u^f \mapsto u^c$ such that

$$(\varphi_i^c, u^c)_{L^2(\Omega)} = \left(\varphi_i^c, u^f\right)_{L^2(\Omega)} \quad \text{for all } i = 1, \dots, n_u^c,$$

or, equivalently,

$$M^c u^c = P^T M^f u^f.$$

We then obtain

$$R = (M^c)^{-1} P^T M^f.$$

Due to P being an exact injection, it follows that $P^T M^f P = M^c$ and thus $RP = \mathbb{I}_{n_u^c}$. Lemma 6.5 then delivers the inverse of $\tilde{G}_u^f - \mathbb{I}_{n_u^f}$ in the form

$$(\tilde{G}_u^f - \mathbb{I}_{n_u^f})^{-1} = P\left[(G_u^c - \mathbb{I}_{n_u^c})^{-1} + \mathbb{I}_{n_u^c}\right] R - \mathbb{I}_{n_u^f},$$

which can be computed by only an inversion of a n_u^c-by-n_u^c matrix from the coarse grid and the inversion of the coarse grid mass matrix in the restriction operator. We obtain an expression for the inverse of the transpose similar to equation (6.13) via

$$\left(((\tilde{G}_u^f)^T - \mathbb{I}_{n_u^f})\right)^{-1} = M\left(\tilde{G}_u^f - \mathbb{I}_{n_u^f}\right)^{-1} M^{-1}.$$

6.3.3 Convergence for classical Newton-Picard

In this section we show that for problem (6.1), LISA (6.2) with classical Newton-Picard preconditioning converges with a grid-independent contraction rate.

For the proof of Theorem 6.7 we need the following lemma. The lemma asserts the existence of a variable transformation which transforms the Hessian blocks to identity, and furthermore reveals the structure of the matrices on the subspaces of fast and slow modes.

Lemma 6.6. *Let* $p \le n_u$, $E_V = \mathrm{diag}(\mu_1, \ldots, \mu_p)$, $E_W = \mathrm{diag}(\mu_{p+1}, \ldots, \mu_{n_u})$. *Then, there exist matrices* $V \in \mathbb{R}^{n_u \times p}$ *and* $W \in \mathbb{R}^{n_u \times (n_u - p)}$ *such that with* $Z = \begin{pmatrix} V & W \end{pmatrix}$ *the following conditions hold:*

(i) Z *is a basis of eigenvectors of* G_u, *i.e.,* $G_u Z = (V E_V \; W E_W)$.
(ii) Z *is* M-*orthonormal, i.e.,* $Z^\mathrm{T} M Z = \mathbb{I}_{n_u}$.
(iii) *There exists a non-singular matrix* T *such that*

$$
T^\mathrm{T} \tilde{J} T = \begin{pmatrix}
\mathbb{I}_{n_u - p} & 0 & 0 & 0 & -\mathbb{I}_{n_u - p} \\
0 & \mathbb{I}_p & 0 & E_V - \mathbb{I}_p & 0 \\
0 & 0 & \gamma N & G_q^\mathrm{T} M V & G_q^\mathrm{T} M W \\
0 & E_V - \mathbb{I}_p & V^\mathrm{T} M G_q & 0 & 0 \\
-\mathbb{I}_{n_u - p} & 0 & W^\mathrm{T} M G_q & 0 & 0
\end{pmatrix},
$$

$$
T^\mathrm{T} \Delta J T = \begin{pmatrix}
0 & 0 & 0 & 0 & -E_W \\
0 & 0 & 0 & 0 & 0 \\
0 & 0 & 0 & 0 & 0 \\
0 & 0 & 0 & 0 & 0 \\
-E_W & 0 & 0 & 0 & 0
\end{pmatrix}.
$$

Proof. The existence of the matrices V and W, as well as conditions (i) and (ii) follow from Lemma 6.1. To show (iii), we choose

$$
T = \begin{pmatrix}
W & V & 0 & 0 & 0 \\
0 & 0 & \mathbb{I}_{n_q m} & 0 & 0 \\
0 & 0 & 0 & V & W
\end{pmatrix}.
$$

Due to M-orthonormality (ii) of V, the Newton-Picard projector from equation (6.12) simplifies to $\Pi = VV^{\mathrm{T}}M$. Using $V^{\mathrm{T}}MW = 0, V^{\mathrm{T}}MV = \mathbb{I}_p$, and $G_u^{\mathrm{T}}MV = MG_uV = MVE_V$ we obtain

$$T^{\mathrm{T}}\Delta JT$$

$$= T^{\mathrm{T}}\begin{pmatrix} 0 & 0 & \left(\Pi^{\mathrm{T}} - \mathbb{I}_{n_u}\right)G_u^{\mathrm{T}}M \\ 0 & 0 & 0 \\ MG_u\left(\Pi - \mathbb{I}_{n_u}\right) & 0 & 0 \end{pmatrix}\begin{pmatrix} W & V & 0 & 0 & 0 \\ 0 & 0 & \mathbb{I}_{n_qm} & 0 & 0 \\ 0 & 0 & 0 & V & W \end{pmatrix}$$

$$= \begin{pmatrix} W^{\mathrm{T}} & 0 & 0 \\ V^{\mathrm{T}} & 0 & 0 \\ 0 & \mathbb{I}_{n_qm} & 0 \\ 0 & 0 & V^{\mathrm{T}} \\ 0 & 0 & W^{\mathrm{T}} \end{pmatrix}\begin{pmatrix} 0 & 0 & 0 & 0 & -MG_uW \\ 0 & 0 & 0 & 0 & 0 \\ -MG_uW & 0 & 0 & 0 & 0 \end{pmatrix}$$

$$= \begin{pmatrix} 0 & 0 & 0 & 0 & -E_W \\ 0 & 0 & 0 & 0 & 0 \\ 0 & 0 & 0 & 0 & 0 \\ 0 & 0 & 0 & 0 & 0 \\ -E_W & 0 & 0 & 0 & 0 \end{pmatrix}.$$

Similarly, we obtain for $\tilde{J}$ the form

$$T^{\mathrm{T}}\tilde{J}T$$

$$= T^{\mathrm{T}}\begin{pmatrix} M & 0 & (MVV^{\mathrm{T}}G_u^{\mathrm{T}} - \mathbb{I}_{n_u})M \\ 0 & \gamma N & G_q^{\mathrm{T}}M \\ M(G_uVV^{\mathrm{T}}M - \mathbb{I}_{n_u}) & MG_q & 0 \end{pmatrix}\begin{pmatrix} W & V & 0 & 0 & 0 \\ 0 & 0 & \mathbb{I}_{n_qm} & 0 & 0 \\ 0 & 0 & 0 & V & W \end{pmatrix}$$

$$= \begin{pmatrix} W^{\mathrm{T}} & 0 & 0 \\ V^{\mathrm{T}} & 0 & 0 \\ 0 & \mathbb{I}_{n_qm} & 0 \\ 0 & 0 & V^{\mathrm{T}} \\ 0 & 0 & W^{\mathrm{T}} \end{pmatrix}\begin{pmatrix} MW & MV & 0 & MV(E_V - \mathbb{I}_p) & -MW \\ 0 & 0 & \gamma N & G_q^{\mathrm{T}}MV & G_q^{\mathrm{T}}MW \\ -MW & MV(E_V - \mathbb{I}_p) & MG_q & 0 & 0 \end{pmatrix}$$

$$= \begin{pmatrix} \mathbb{I}_{n_u-p} & 0 & 0 & 0 & -\mathbb{I}_{n_u-p} \\ 0 & \mathbb{I}_p & 0 & E_V - \mathbb{I}_p & 0 \\ 0 & 0 & \gamma N & G_q^{\mathrm{T}}MV & G_q^{\mathrm{T}}MW \\ 0 & E_V - \mathbb{I}_p & V^{\mathrm{T}}MG_q & 0 & 0 \\ -\mathbb{I}_{n_u-p} & 0 & W^{\mathrm{T}}MG_q & 0 & 0 \end{pmatrix}. \quad \Box$$

We now state the central theorem of this section.

Theorem 6.7. *Let μ_i, $i = 1, \ldots, n_u$, denote the eigenvalues of G_u ordered in descending modulus, let $1 < p \leq n_u$, and assume $\mu_p > \mu_{p+1}$. We further assume the existence of a linear operator $\bar{G}_u : L_2(\Sigma) \to L_2(\Omega)$ which is continuous, i.e.,*

$$\left\| \bar{G}_q q \right\|_{L_2(\Omega)} \leq C_1 \|q\|_{L_2(\Sigma)} \quad \text{for all } q \in L_2(\Sigma), \tag{6.14}$$

and satisfies the discretization error condition

$$\left\| \sum_{j=1}^{n_u} (G_q q)_j \varphi_j - \bar{G}_q q \right\|_{L_2(\Omega)} \leq C_2 \|q\|_{L_2(\Sigma)} \quad \text{for all } q \in \mathbb{R}^{n_q m}, q = \sum_{i=1}^{n_q m} q_i \psi_i, \tag{6.15}$$

with constants $C_1, C_2 \in \mathbb{R}$. If

$$\gamma > (C_1 + C_2)^2 / (1 - \mu_1)^2$$

then LISA (6.2) with Newton-Picard preconditioning applied to problem (6.1) converges with a contraction rate of at most μ_{p+1}/μ_1.

Proof. Due to Theorem 5.26 the contraction rate is given by the spectral radius $\sigma_{\mathrm{r}}(\tilde{J}^{-1}\Delta J) = \sigma_{\mathrm{r}}(T^{-1}\tilde{J}^{-1}T^{-\mathrm{T}}T^{\mathrm{T}}\Delta JT)$. We obtain the eigenvalue problem

$$\left(T^{\mathrm{T}}\tilde{J}T \right)^{-1} T^{\mathrm{T}}\Delta JTv - \sigma v = 0,$$

which is equivalent to solving the generalized eigenvalue problem

$$-T^{\mathrm{T}}\Delta JTv + \sigma T^{\mathrm{T}}\tilde{J}Tv = 0,$$

with the matrices given by Lemma 6.6. We prove the theorem by contradiction. Assume that there is a complex eigenpair (v, σ) such that $|\sigma| \geq \mu_{p+1}/\mu_1 > \mu_{p+1}$. Division by σ yields the system

$$(1/\sigma)E_W v_5 + v_1 - v_5 = 0, \tag{6.16a}$$

$$v_2 + (E_V - \mathbb{I}_p) v_4 = 0, \tag{6.16b}$$

$$\gamma N v_3 + G_q^{\mathrm{T}} M (V v_4 + W v_5) = 0, \tag{6.16c}$$

$$(E_V - \mathbb{I}_p) v_2 + V^{\mathrm{T}} M G_q v_3 = 0, \tag{6.16d}$$

$$(1/\sigma)E_W v_1 - v_1 + W^{\mathrm{T}} M G_q v_3 = 0, \tag{6.16e}$$

where v was divided into five parts $v_1, \ldots, v_5$ corresponding to the blocks of the system. Because $|\sigma| > \mu_{p+1}$ we obtain invertibility of $\mathbb{I}_{n_u-p} - (1/\sigma)E_W$ and thus we can eliminate

$$v_5 = \left(\mathbb{I}_{n_u-p} - (1/\sigma)E_W\right)^{-1} v_1, \quad v_4 = \left(\mathbb{I}_p - E_V\right)^{-1} v_2, \tag{6.17a}$$

$$v_2 = \left(\mathbb{I}_p - E_V\right)^{-1} V^{\mathsf{T}} M G_q v_3, \quad v_1 = \left(\mathbb{I}_{n_u-p} - (1/\sigma)E_W\right)^{-1} V^{\mathsf{T}} M G_q v_3. \tag{6.17b}$$

Substituting these back in equation (6.16c) yields

$$\left(\gamma N + G_q^{\mathsf{T}} M V \left(\mathbb{I}_p - E_V\right)^{-2} V^{\mathsf{T}} M G_q \right.$$
$$\left. + G_q^{\mathsf{T}} M W \left(\mathbb{I}_{n_u-p} - \sigma^{-1} E_W\right)^{-2} W^{\mathsf{T}} M G_q\right) v_3 = 0.$$

We denote the complex valued matrix on the left hand side with $A(\sigma)$. The final step of the proof consists of showing that $A(\sigma)$ is invertible if γ is large enough. Since M and N are positive definite matrices they have Cholesky decompositions

$$M = R_M^{\mathsf{T}} R_M, \quad N = R_N^{\mathsf{T}} R_N$$

with invertible $R_M \in \mathbb{R}^{n_u \times n_u}, R_N \in \mathbb{R}^{n_q m \times n_q m}$. If we define

$$B(\sigma) := \begin{pmatrix} \mathbb{I}_p - E_V & 0 \\ 0 & \mathbb{I}_{n_u-p} - \sigma^{-1} E_W \end{pmatrix}^{-1} \begin{pmatrix} V \\ W \end{pmatrix} R_M^{\mathsf{T}} R_M G_q R_N^{-1} \in \mathbb{C}^{n_q m \times n_q m}$$

we obtain

$$\gamma^{-1} R_N^{-\mathsf{T}} A(\sigma) R_N^{-1} = \mathbb{I}_{n_q m} + \gamma^{-1} B(\sigma)^{\mathsf{T}} B(\sigma).$$

We now estimate the two-norm

$$\left\| B(\sigma)^{\mathsf{T}} B(\sigma) \right\|_2$$

$$\leq \left\| R_M G_q R_N^{-1} \right\|_2^2 \left\| R_M (V\ W) \right\|_2^2 \left\| \begin{pmatrix} \mathbb{I}_p - E_V & 0 \\ 0 & \mathbb{I}_{n_u-p} - \sigma^{-1} E_W \end{pmatrix}^{-1} \right\|_2^2. \tag{6.18}$$

We consider each of the norms on the right hand side of inequality (6.18) separately. Due to Lemma 6.6 (ii) we obtain

$$\left\| R_M (V\ W) \right\|_2^2 = 1.$$

The matrix in the last term of inequality (6.18) is a diagonal matrix and so the maximum singular value of it can be easily determined. Due to $|\sigma| > \mu_{p+1}/\mu_1$ we have that

$$\left\| \begin{pmatrix} \mathbb{I}_p - E_V & 0 \\ 0 & \mathbb{I}_{n_u-p} - \sigma^{-1} E_W \end{pmatrix}^{-1} \right\|_2 \leq \frac{1}{1 - \mu_1}.$$

The first term of inequality (6.18) can be bounded considering

$$
\begin{aligned}
\left\| R_M G_q R_N^{-1} \right\|_2 &= \sup_{\substack{\|R_N q\|_2 = 1 \\ q \in \mathbb{R}^{n_q m}}} \left\| R_M G_q q \right\|_2 \\
&= \sup_{\substack{q = \sum_{i=1}^{n_q m} q_i \psi_i \\ \iint_\Sigma q^2 = 1}} \left(\int_\Omega \left(\sum_{j=1}^{n_u} (G_q q)_j \varphi_j \right)^2 \right)^{\frac{1}{2}} \\
&= \sup_{\substack{q = \sum_{i=1}^{n_q m} q_i \psi_i \\ \|q\|_{L_2(\Sigma)} = 1}} \left\| \sum_{j=1}^{n_u} (G_q q)_j \varphi_j - \bar{G}_q q + \bar{G}_q q \right\|_{L_2(\Omega)} \\
&\le C_2 + \sup_{\substack{q = \sum_{i=1}^{n_q m} q_i \psi_i \\ \|q\|_{L_2(\Sigma)} = 1}} \left\| \bar{G}_q q \right\|_{L_2(\Omega)} \\
&\le C_2 + \sup_{\substack{q \in L_2(\Sigma) \\ \|q\|_{L_2(\Sigma)} = 1}} \left\| \bar{G}_q q \right\|_{L_2(\Omega)} \\
&\le C_1 + C_2.
\end{aligned}
$$

If now $\gamma > (C_1 + C_2)^2 / (1 - \mu_1)^2$ then $\left\| \gamma^{-1} B(\sigma)^{\mathrm{T}} B(\sigma) \right\|_2 < 1$ and thus $A(\sigma)$ is invertible. It follows that $v_3 = 0$, which implies $v = 0$ via equations (6.17). Thus, (v, σ) cannot be an eigenpair. $\square$

The main result of this section is now at hand:

Corollary 6.8. *The asymptotic convergence rate of LISA with classical Newton-Picard preconditioning on the model problem is mesh-independent, provided γ is large enough.*

Proof. For finer and finer discretizations the largest $p + 1$ eigenvalues of $M^{-1}L$ converge. Thus, also the eigenvalues μ_1 and μ_{p+1} of G_u converge to some $\bar{\mu}_1 < \bar{\mu}$ and $\bar{\mu}_{p+1} \le \bar{\mu}_1$, with $\bar{\mu}$ given by Lemma 6.3. We construct $\bar{G}_q$ as the infinite dimensional counterpart to G_q, i.e., $\bar{G}_q$ maps controls in $L_2(\Sigma)$ to the end value of the heat equation (6.1b)–(6.1c) with zero initial values for the state. This operator is continuous (see, e.g., Tröltzsch [152]). Let $\varepsilon > 0$. We can assume that G_q satisfies the discretization error condition (6.15) with $C_2 = \varepsilon$ for a fine enough and also for all finer discretizations. Define

$$
\bar{\gamma} = (C_1 + \varepsilon)^2 / (1 - \bar{\mu})^2.
$$

Theorem 6.7 yields that if $\gamma > \bar{\gamma}$ then the asymptotic convergence rate of LISA is below the mesh-independent bound $\bar{\mu}_{p+1} / \bar{\mu}_1$. $\square$

We remark here that our numerical experience suggests the conjecture that the contraction rate is actually μ_{p+1} instead of μ_{p+1}/μ_1.

6.3.4 Numerical solution of the approximated linear system

The implicit inversion of the preconditioner $\tilde{J}$ can be carried out by block elimination. To simplify notation we denote the residuals by $r_i, i = 1, 2, 3$. We want to solve

$$
\begin{pmatrix}
M & 0 & (\tilde{G}_u^{\mathrm{T}} - \mathbb{I}_{n_u})M \\
0 & \gamma N & G_q^{\mathrm{T}} M \\
M(\tilde{G}_u - \mathbb{I}_{n_u}) & M G_q & 0
\end{pmatrix}
\begin{pmatrix}
u_0 \\ q \\ \lambda
\end{pmatrix}
=
\begin{pmatrix}
r_1 \\ r_2 \\ r_3
\end{pmatrix}.
$$

Solving the last block-row for u_0 and the first for λ, the second block-row becomes

$$
Hq = r_2 - G_q^{\mathrm{T}}\left(\tilde{G}_u^{\mathrm{T}} - \mathbb{I}_{n_u}\right)^{-1}\left(r_1 - M\left(\tilde{G}_u - \mathbb{I}_{n_u}\right)^{-1} M^{-1} r_3\right) =: \tilde{r}, \qquad (6.19)
$$

with the $n_q m$-by-$n_q m$ symmetric positive-definite matrix

$$
H = \gamma N + G_q^{\mathrm{T}}\left(\tilde{G}_u^{\mathrm{T}} - \mathbb{I}_{n_u}\right)^{-1} M\left(\tilde{G}_u - \mathbb{I}_{n_u}\right)^{-1} G_q.
$$

If $n_q m$ is moderately small we can set up G_q as well as $(\tilde{G}_u - \mathbb{I}_{n_u})^{-1} G_q$ according to Lemma 6.5 or the two-grid analog and thus form matrix H explicitly. Then, equation (6.19) can be solved for q via Cholesky decomposition of H. Alternatively, we can employ a Preconditioned Conjugate Gradient (PCG) method with preconditioner N.

Lemma 6.9. *Assume there exists a linear operator* $\bar{G}_u : L_2(\Sigma) \to L_2(\Omega)$ *which satisfies assumptions (6.14) and (6.15). Then the spectral condition number of matrix* $N^{-1}H$ *is bounded by*

$$
\mathrm{cond}_2(N^{-1}H) \le 1 + \frac{(C_1 + C_2)^2}{\gamma(1 - \bar{\mu})^2},
$$

with $\bar{\mu}$ *from Lemma 6.3.*

Proof. The spectral condition number of $N^{-1}H$ is equal to the ratio of largest to smallest eigenvalue of $N^{-1}H$. Let $(q, \sigma) \in \mathbb{R}^{n_q m} \times \mathbb{R}$ be an eigenpair of $N^{-1}H$, i.e.,

$$
Hq - \sigma N q = 0
$$

and define $q = \sum_{i=1}^{n_q m} q_i \psi_i \in L_2(\Sigma)$. We obtain

$$\sigma q^T N q = q^T H q = \gamma q^T N q + \left\| R_M (\tilde{G}_u - \mathbb{I}_{n_u})^{-1} R_M^{-1} R_M G_q q \right\|_2^2 \tag{6.20a}$$

$$\begin{cases} \geq \gamma q^T N q & \Rightarrow \quad \sigma \geq \gamma, \\ \leq \gamma q^T N q + \left\| R_M (\tilde{G}_u - \mathbb{I}_{n_u})^{-1} R_M^{-1} \right\|_2^2 \left\| R_M G_q q \right\|_2^2, \end{cases} \tag{6.20b}$$

By virtue of Lemma 6.3 the largest singular value of $\tilde{G}_u - \mathbb{I}_{n_u}$ is bounded by $1 - \bar{\mu}$ and thus we obtain

$$\left\| R_M (\tilde{G}_u - \mathbb{I}_{n_u})^{-1} R_M^{-1} \right\|_2^2 \leq 1/(1 - \bar{\mu})^2. \tag{6.21}$$

For the remaining norm we consider

$$\left\| R_M G_q q \right\|_2 = \left\| \sum_{i=1}^{n_q m} (G_q q)_i \psi_i \right\|_{L^2(\Sigma)} \tag{6.22a}$$

$$\leq \left\| \sum_{i=1}^{n_q m} (G_q q)_i \psi_i - \bar{G}_q q \right\|_{L^2(\Sigma)} + \left\| \bar{G}_q q \right\|_{L^2(\Sigma)} \tag{6.22b}$$

$$\leq (C_1 + C_2) \left\| q \right\|_{L^2(\Sigma)} = (C_1 + C_2) \left\| R_N q \right\|_2. \tag{6.22c}$$

We now combine inequalities (6.20), (6.21), and (6.22) to obtain the assertion. $\square$

As a consequence of Lemma 6.9 we obtain that the number of required PCG iterations bounded by a grid-independent number. In our numerical experience 10–20 PCG iterations usually suffice for a reduction of the relative residual to 10^{-6}.

Solving for u_0 and λ is then simple:

$$u_0 = \left(\tilde{G}_u - \mathbb{I}_{n_u} \right)^{-1} \left(M^{-1} r_3 - G_q q \right), \lambda = M^{-1} \left(\tilde{G}_u^T - \mathbb{I}_{n_u} \right)^{-1} (r_1 - M u_0).$$

Note that once G_q and $\tilde{G}_u$ (in a suitable representation) have been calculated no further numerical integration of the system dynamics is required.

6.3.5 Pseudocode

In this section we provide pseudocode to sketch the implementation of the proposed Newton-Picard preconditioners. We focus on the case $n_q m \ll n_u$ in which it is economical to solve equation (6.19) by forming H and using Cholesky decomposition, but we also discuss implementation alternatives for the case of large $n_q m$. We use a Matlab® oriented syntax here and assume that the reader is familiar with linear algebra routines in Matlab®. For readability purposes we further assume that matrices $L, M, L^c, M^c, P, R, E, X, N$ and dimensions n_u, n_q, m are globally accessible in each function. We do not discuss the assembly of Galerkin matrices L, M, L^c, M^c, and U or grid transfer operators P and R here.

The first function computes the matrices needed to evaluate the classical Newton-Picard preconditioner later. For numerical stability it is advantageous to perform IRAM in `eigs` with the symmetrified version (see proof of Lemma 6.1) and to explicitly set the option that the matrix is real symmetric (not shown in pseudocode).

Function $[P, R, E, X]$ = `classicalApprox`

---

output: Matrices $P \in \mathbb{R}^{n_u \times p}, R \in \mathbb{R}^{p \times n_u}$ with $RP = \mathbb{I}_p$, matrices $E, X \in \mathbb{R}^{p \times p}$

$R_M = \text{chol}(M)$;
$[\tilde{V}, \tilde{E}] = \text{eigs}(\mathbb{Q}(u) \, R_M^{\mathrm{T}} \setminus (L * (R_M \setminus u)), \, n_u, \, p, \, 'la')$;
$P = R_M \setminus \tilde{V}$;
$R = P^{\mathrm{T}} * M$;
$E = \text{diag}(\exp(\text{diag}(\tilde{E})))$;
$X = \text{diag}(1./(\text{diag}(E) - 1) + 1)$;

For the two-grid version, we assume that the prolongation P and restriction R are given. Because the occurring matrices are small, we can employ the LAPACK methods in `eig` instead of IRAM in `eigs`.

Function $[E, X]$ = `coarseGridApprox`

---

output: Coarse grid approximation $E \in \mathbb{R}^{n_u^c \times n_u^c}$ of G_u, matrix $X \in \mathbb{R}^{n_u^c \times n_u^c}$

$[\tilde{V}, \tilde{E}] = \text{eig}(\text{full}(L^c), \text{full}(M^c))$;
$E = \tilde{V} * \text{diag}(\exp(\text{diag}(\tilde{E}))) \, / \, \tilde{V}$;
$X = \text{inv}(E - \mathbb{I}_{n_u^c}) + \mathbb{I}_{n_u^c}$;

Now P, R, E, X are known and we can formulate matrix vector and matrix transpose vector products with $\tilde{G}_u$.

Function u_1 = `Gup`(u_0)	**Function u_0** = `GupT`(u_1)
input : $u_0 \in \mathbb{R}^{n_u}$	**input** : $u_1 \in \mathbb{R}^{n_u}$
output: $u_1 = \tilde{G}_u u_0 \in \mathbb{R}^{n_u}$	**output**: $u_0 = \tilde{G}_u^{\mathrm{T}} u_1 \in \mathbb{R}^{n_u}$
$u_1 = (P * (E * (R * u_0)))$;	$u_0 = (R^{\mathrm{T}} * (E^{\mathrm{T}} * (P^{\mathrm{T}} * u_1)))$;

In the same way we can evaluate matrix vector and matrix transpose vector products with the inverse of $\tilde{G}_u - \mathbb{I}_{n_u}$ according to Lemma 6.5.

Function u = `iGupmI`(r)	**Function** r = `iGupTmI`(u)
input : $r \in \mathbb{R}^{n_u}$ **output**: $u = (\tilde{G}_u - \mathbb{I}_{n_u})^{-1} r$	**input** : $u_1 \in \mathbb{R}^{n_u}$ **output**: $r = (\tilde{G}_u^{\mathrm{T}} - \mathbb{I}_{n_u})^{-1} u$
$u = (P * (X * (R * r))) - r;$	$r = (R^{\mathrm{T}} * (X^{\mathrm{T}} * (P^{\mathrm{T}} * u))) - u;$

We want to remark that we take the liberty to call the four previous functions also with matrix arguments. In this case the respective function is understood to return a matrix of the same size and to be evaluated on each column of the input matrix. For the computation of matrix vector products with G_u and G_q we define an auxiliary function which integrates ODE (6.4) for given initial state and control variables. The control coefficients are constant in time.

Function u_{e} = `dG`$(\Delta t, u_{\mathrm{s}}, \tilde{q})$
input : Duration Δt, initial value $u_{\mathrm{s}} \in \mathbb{R}^{n_u}$, control coefficients $\tilde{q} \in \mathbb{R}^{n_q}$ **output**: End state $u_{\mathrm{e}} \in \mathbb{R}^{n_u}$ after time Δt
Solve ODE (6.4) with initial value u_{s} and constant control $\tilde{q}$, e.g., by `ode15s`;

Based on the previous function we can now assemble matrix G_q. There are alternative ways for the assembly. We have chosen an approach for the case that large intervals for `dG` can be efficiently and accurately computed through adaptive step size control as in, e.g., `ode15s`.

Function G_q = `computeGq`
output: Matrix $G_q \in \mathbb{R}^{n_u \times n_q m}$

for $j = 1 : n_q$ **do**
 $\quad\lfloor\ G_q(:, j + n_q * (m-1)) = \mathrm{dG}(1\,/\,m, 0, \mathbb{I}_{n_q}(:, j));$
 for $i = 1 : m - 1$ **do**
 $\quad$ **for** $j = 1 : n_q$ **do**
 $\quad\quad\lfloor\ G_q(:, j + n_q * (i-1)) = \mathrm{dG}(1 - i\,/\,m, G_q(:, j + n_q * (m-1)), 0);$

We can alternatively compute matrix vector and matrix transpose vector products with G_q via the following functions. For the transpose we exploit the expression

$$\left(\partial G_u^i \partial G_q\right)^{\mathrm{T}} = U^{\mathrm{T}} L^{-1} \left(\partial G_u^{\mathrm{T}} - \mathbb{I}_{n_u}\right) \left(\partial G_u^{\mathrm{T}}\right)^i = U^{\mathrm{T}} L^{-1} M \left(\partial G_u^{i+1} - \partial G_u^i\right) M^{-1}.$$

Function u_1 = Gq(q)

input : $q \in \mathbb{R}^{n_q m}$
output: $u_1 = G_q q \in \mathbb{R}^{n_u}$

$u_1 = \texttt{zeros}(n_u, 1)$;
for $i = 0 : m - 1$ **do**
$\quad \lfloor \; u_1 = \texttt{dG}(1 \; / \; m, \; u_1, \; q(i * n_q + (1 : n_q)))$;

Function q = GqT(λ)

input : $\lambda \in \mathbb{R}^{n_u}$
output: $q = G_q^{\mathrm{T}} \lambda \in \mathbb{R}^{n_q m}$

$q = \texttt{zeros}(n_q * m, 1)$; $\tilde{\lambda}^+ = M \setminus \lambda$;
for $i = m - 1 : -1 : 0$ **do**
$\quad \lvert \; \tilde{\lambda} = \tilde{\lambda}^+$; $\tilde{\lambda}^+ = \texttt{dG}(1 \; / \; m, \; \tilde{\lambda}, \; 0)$;
$\quad \lfloor \; q(i * n_q + (1 : n_q)) = U^{\mathrm{T}} * (L \setminus (M * (\tilde{\lambda}^+ - \tilde{\lambda})))$;

We can also formulate functions for matrix vector and matrix transpose vector products with G_u.

Function u_1 = Gu(u_0)

input : $u_0 \in \mathbb{R}^{n_u}$
output: $u_1 = G_u u_0 \in \mathbb{R}^{n_u}$

$u_1 = \texttt{dG}(1, u_0, 0)$;

Function u_0 = GuT(λ)

input : $\lambda \in \mathbb{R}^{n_u}$
output: $u_0 = G_u^{\mathrm{T}} \lambda \in \mathbb{R}^{n_u}$

$u_0 = M * \texttt{dG}(1, M \setminus \lambda, 0)$;

For the evaluation of the preconditioner we employ a Cholesky decomposition of matrix H which can be obtained with the following function.

Function R_H = decompH

output: Cholesky factor $R_H \in \mathbb{R}^{n_q m \times n_q m}$ of $H = R_H^{\mathrm{T}} R_H$

$V = \texttt{iGupmI}(G_q)$;
$R_H = \texttt{chol}(\gamma * N + V^{\mathrm{T}} * M * V)$;

We can finally state the function for a matrix vector product with the symmetric indefinite matrix $\hat{J}$. For readability we split up the argument and result into three subvectors.

Function $[r_1, r_2, r_3]$ = J(u, q, λ)

> **input** : $u \in \mathbb{R}^{n_u}, q \in \mathbb{R}^{n_q m}, \lambda \in \mathbb{R}^{n_u}$
> **output:** $[r_1; r_2; r_3] = J[u; q; \lambda] \in \mathbb{R}^{n_u + n_q m + n_u}$
>
> $r_1 = M * u + \text{GuT}(\lambda) - \lambda;$
> $r_2 = \gamma * N * q + G_q^{\text{T}} * \lambda;$
> $r_3 = \text{Gu}(u) - u + G_q * q;$

At last we present pseudocode for matrix vector products with the preconditioner $\tilde{J}^{-1}$. Again, we split up the argument and result into three subvectors.

Function $[u, q, \lambda]$ = iJp(r_1, r_2, r_3)

> **input** : $r_1 \in \mathbb{R}^{n_u}, r_2 \in \mathbb{R}^{n_q m}, r_3 \in \mathbb{R}^{n_u}$
> **output:** $[u; q; \lambda] = \tilde{J}^{-1}[r_1; r_2; r_3] \in \mathbb{R}^{n_u + n_q m + n_u}$
>
> $q = R_H \setminus (R_H^{\text{T}} \setminus (r_2 - \text{iGupTmI}(G_q^{\text{T}} * (r_1 - M * \text{iGupmI}(r_3)))));$
> $u = \text{iGupmI}(r_3 - G_q * q);$
> $\lambda = \text{iGupTmI}(r_1 - M * u);$

We can also substitute the functions Gq and GqT for the occurrences of G_q and G_q^{T} in J and iJp.

6.3.6 Algorithmic complexity

In this section we discuss the algorithmic complexity of the proposed method. To simplify analysis we only count the number of necessary (fine grid) system integrations which are required when evaluating matrix vector or matrix transpose vector products with G_u, G_q, or $(G_u \ G_q)$. We shall see that we need $\mathscr{O}(\|\hat{J}\hat{z}^0 + \hat{F}\| / \varepsilon_0)$ simulations to solve the optimization problem (4.1) up to an absolute tolerance of $\varepsilon_0 > 0$.

If we solve the reduced systems (6.19) exactly then we obtain a grid-independent contraction bound $\kappa = \sigma_{\text{r}}(\tilde{J}^{-1}\Delta J)$ by virtue of Corollary 6.8. By Lemma 6.9 we know that we can solve the reduced system (6.19) up to a relative residual tolerance $\varepsilon_H > 0$ using PCG with a grid-independently bounded number k of iterations. A backward analysis in the sense of Lemma 5.1 yields a matrix $\tilde{H}$ such that $\|H - \tilde{H}\| \leq \|\tilde{r}\| \varepsilon_H$ and such that the PCG iterate q^k satisfies $\tilde{H}q^k = \tilde{r}$. Additionally, matrix vector products with the inverse mass matrix M^{-1} need to be evaluated. This can also be done at linear cost using diagonal preconditioners with PCG (see Wathen [162]) to tolerance $\varepsilon_M > 0$. A similar backward analysis as for $\tilde{H}$ yields a perturbed mass matrix $\tilde{M}$. Because the eigenvalues of the now $\tilde{H}$ and $\tilde{M}$ dependent iteration matrix (as a perturbation of $\tilde{J}^{-1}\Delta J$) depend continuously on the entries of $\tilde{H}$ and $\tilde{M}$ we obtain that for each $\tilde{\kappa} \in (\kappa, 1)$ there exist $\varepsilon_H, \varepsilon_M > 0$ such that the contraction rate of the outer iteration is bounded by $\tilde{\kappa}$. Thus, we can solve the optimization problem (4.1) up to a tolerance $\varepsilon_O > 0$ within $\mathcal{O}(\|\hat{J}_z^0 + \hat{F}\|/\varepsilon_O)$ iterations.

We now count the number of system integrations per iteration under the assumption that we perform n_H inner PCG iterations per outer iteration. For the evaluation of matrix vector products with $\hat{J}$ we need two system integrations for maxtrix vector products with $(G_u\ G_q)$ and its transpose. Concerning the matrix vector product with the preconditioner $\tilde{J}^{-1}$ we observe that multiplications with $\tilde{G}_u$ and $(\tilde{G}_u - \mathbb{I}_{n_u})^{-1}$ do not need any system integrations. However, the setup of $\tilde{r}$ in equation (6.19) requires one integration for a matrix vector product with G_q. Furthermore, each inner PCG iteration requires two additional simulations for matrix vector products with G_q and its transpose. Thus we need $3 + 2n_H$ system simulations per outer iteration which yields an optimal complexity of $\mathcal{O}(1/\varepsilon_O)$ system integrations for the solution of the optimization problem (4.1) up to tolerance ε_O.

When performed this way the additional linear algebra consists of matrix vector multiplications with sparse matrices, lower order vector computations, and dense linear algebra for system sizes bounded by a factor of the grid-independent numbers p or n_u^c, respectively.

In the case of classical Newton-Picard approximation we need to add the complexity of IRAM for the one-time determination of the dominant subspace spanned by V. A detailed analysis of the numerical complexity for this step is beyond the scope of this thesis. Suffice it that based on Saad [135, Theorem 6.3 and Chebyshev polynomial approximation (4.49)] together with the approximation $\mu_{n_u} \approx 0$ we assume that the tangent of the angle between the p-th eigenvector of G_u and the l-th Krylov subspace decreases linearly in l with a factor depending on the ultimately grid-independent ratio μ_p/μ_{p+1} which we assume to be greater than one. We need one matrix vector product with G_u per Arnoldi iteration. Because this computation is only needed once independently of ε_O the asymptotic complexity $\mathcal{O}(1/\varepsilon_O)$ does not deteriorate. It does, however, have an effect for practical computations (see Section 12.3) and can easily dominate the overall cost of the algorithm already for modest values of p.

We have also found the approach with explicit solution of equation (6.19) via Cholesky decomposition of H beneficial for the practical computations presented in Chapter 12. Although we obtain a one-time cubic complexity in $n_q m$ and a square complexity in $n_q m$ per outer iteration, the runtime can be much faster than iterative solution of equation (6.19) because per outer iteration only two system integrations are required instead of $3 + 2n_H$.

We want to close this section with the remark that the optimal choice of p, $\tilde{\kappa}$, ε_M, and ε_H is a complex optimization problem which exceeds the scope of this thesis.

6.4 Extension to nonlinear problems and Multiple Shooting

So far we have focused our investigation of Newton-Picard preconditioning on the linear model problem (6.1). For nonlinear problems we have to deal with the difficulty that the matrix G_u depends on the current iterate and that thus the dominant subspace can change from one SQP iteration to another. The extension of the two-grid Newton-Picard preconditioner to nonlinear problems is straight-forward because the dominant subspace is implicitly given by the coarse grid approximation. For the classical Newton-Picard preconditioner, however, the situation is more complicated.

Lust et al. [110] use a variant of the Subspace Iteration, originally developed by Stewart [149], to update the basis V of the dominant subspace approximation in each SQP iteration. The Subspace Iteration is an iterative method for the computation of eigenvalues and eigenvectors. Each iteration consists of three steps (see, e.g., Saad [133]):

1. Compute $V := G_u V$.
2. Orthonormalize V.
3. Use the QR algorithm on $V^T G_u V$ to compute its Schur vectors Y and update $V := VY$ (Schur-Rayleigh-Ritz step).

Locking and shifting techniques can improve efficiency of the method in practical implementations (see Saad [133]).

The Subspace Iteration is also used simultaneously such that only a few Subspace Iterations (ideally only one) are needed per SQP step. Potschka et al. [130] present preliminary numerical results for a Newton-Picard inexact SQP method without LISA and using the Euclidean projector.

The reader has surely noticed that we have so far in this chapter only considered the case of Single but not Multiple Shooting. Again, the two-grid preconditioner can be extended in a rather canonical way (see Chapter 8 for the remaining details). For the classical Newton-Picard preconditioner we can sketch two approaches:

Sequential approach. We perform the Subspace Iteration on the product of the local shooting matrices $G_u := G_u^{n_{MS}} \ldots G_u^1$. The main drawback of the sequential approach is that it is impossible to compute $G_u V$ in parallel because the result of $G_u^1 V$ must be available to compute $G_u^2 G_u^1 V$ and so forth.

Simultaneous approach. We introduce a local dominant subspace approximation basis V^i on each shooting interval and perform the Subspace Iterations in a decoupled way. It is, however, at least unclear how the local error propagates to the accumulated error in the product because $G_u^i V^i$ and V^{i+1} will in general not span the same subspace. Furthermore, the convergence speed of the Subspace Iteration decreases with shorter time intervals, which can be seen for the linear model problem by considering the matrix exponential for G_u^i on a shooting interval of length τ. We obtain for the eigenvalues that $\mu_j^i = \exp(\tau \tilde{\mu}_j^i)$. Thus for smaller τ we obtain larger eigenvalue moduli and ratios which impair the convergence speed of the Subspace Iteration on each interval.

Based on these considerations we have decided to develop only the two-grid version fully for nonlinear problems and Multiple Shooting. For an appropriate Hessian approximation for nonlinear problems we also refer the reader to Chapter 8.

7 One-shot one-step methods and their limitations

We want to address the basic question if the results of Chapter 6 for the convergence of the Newton-Picard LISA can be extended to general one-shot one-step methods. We shall explain this class of problems and see that in the general case no such result as Theorem 6.7 for the model problem (6.1) is possible. For completeness we quote large passages of the article Bock et al. [28] with modifications concerning references to other parts of this thesis.

Many nonlinear problems

$$g(x_s, x_c) = 0, \quad x = (x_s, x_c) \in \mathbb{R}^{m+(n-m)}, g \in \mathscr{C}^1(\mathbb{R}^n, \mathbb{R}^m), \tag{7.1}$$

with fixed x_c can be successfully solved with Newton-type methods (see Chapter 5)

$$\text{given } x_s^0, \quad x_s^{k+1} = x_s^k - G_k^{-1} g(x_s^k, x_c). \tag{7.2}$$

In most cases a cheap approximation $G_k \approx \frac{\partial g}{\partial x_s}(x_s^k, x_c)$ with linear contraction rate of, say, $\kappa = 0.8$ is already good enough to produce an efficient numerical method. In general, cheaper computation of the action of G_k^{-1} on the residual compared to the action of $(\frac{\partial g}{\partial x_s})^{-1}$ must compensate for the loss of locally quadratic convergence of a Newton method to obtain an overall performance gain within the desired accuracy. It is a tempting idea to use the same Jacobian approximations G_k from the Newton-type method in an inexact SQP method for the optimization problem with the same constraint

$$\min_{x \in \mathbb{R}^n} f(x) \quad \text{s.t. } g(x) = 0. \tag{7.3}$$

From this point of view we call problem (7.1) the *forward problem* of optimization problem (7.3) and we will refer to the variables x_c as *control* or *design* variables and to x_s as *state* variables.

Using (inexact) SQP methods which do not satisfy $g = 0$ in every iteration for (7.3) is usually called *simultaneous*, or *all-at-once* approach and has proved to be successful for several applications, e.g., in aerodynamic shape optimization Bock et al. [27], Hazra et al. [80], chemical engineering Potschka et al. [130], or for the model problem (6.1) in Chapter 6. Any inexact SQP method for equality constrained problems of the form (7.3) is equivalent to a Newton-type method for the necessary optimality conditions

$$\nabla_x L(x,y) = 0, \quad g(x) = 0,$$

as we have seen in Chapters 4 and 5. We are lead to a Newton-type iteration for the primal-dual variables $z = (x,y) \in \mathbb{R}^{n+m}$

$$z^{k+1} = z^k - \begin{pmatrix} H_k & A_k^T \\ A_k & 0 \end{pmatrix}^{-1} \begin{pmatrix} \nabla_z L(z^k) \\ g(x^k) \end{pmatrix}, \tag{7.4}$$

where H_k is an approximation of the Hessian of the Lagrangian L and A_k is an approximation of the constraint Jacobian

$$\frac{dg}{dx} \approx A_k = \begin{pmatrix} A_{1k} & A_{2k} \end{pmatrix}.$$

Note that like in Chapter 6 we use a plus sign in the definition of the Lagrangian here to obtain symmetry of the KKT matrices. If $A_{1k} = G_k$ holds the method is called *one-step* because exactly one step of the solver for the forward and the adjoint problem is performed per optimization iteration.

The discretized model problem (6.1) in Chapter 6 has exactly this structure where A_k is implicitly given by a Newton-Picard preconditioner for the forward problem of finding a periodic steady state. Theorem 6.7 shows that in the case of the model problem we achieve almost the same contraction for the optimization problem by simply reusing G_k in equation (7.4).

In the remainder of this chapter we illustrate with examples that in general only little connection exists between the convergence of Newton-type methods (7.2) for the forward problem and the convergence of simultaneous one-step inexact SQP methods (7.4) for the optimization problem because the coupling of control, state, and dual variables gives rise to an intricate feedback between each other within the optimization problem.

Griewank [71] discusses that in order to guarantee convergence of the simultaneous optimization method this feedback must be broken up, e.g., by keeping the design y fixed for several optimization steps, or by at least damping the feedback in the update of the design y by the use of "preconditioners" for which he derives a necessary condition for convergence based on an eigenvalue analysis.

We are interested in the different but important case where there exists a contractive method for the forward problem (e.g., Bock et al. [27], Hazra et al. [80], Potschka et al. [130], and Chapter 6). If applied to the linearized forward problem, we obtain preconditioners which are contractive, i.e., the eigenvalues of the pre-conditioned system lie in a ball around 1 with radius less than 1. The contraction property suggests the use of a simultaneous one-step approach. However, we can show that contraction for the forward problem is neither sufficient nor necessary for convergence of the simultaneous one-step method.

The structure of this chapter is the following: Based on the Local Contraction Theorem 5.5 we present in Section 7.1 illustrative, counter-intuitive examples of convergence and divergence for the forward and optimization problem which form the basis for the later investigations on recovery of convergence. We continue with presenting a third example and three prototypical subproblem regularization strategies in Section 7.2 and perform an asymptotic analysis for large regularization parameters in Section 7.3. We also show de facto loss of convergence for one of the examples and compare the regularization approaches to Griewank's One-Step One-Shot projected Hessian preconditioners.

7.1 Illustrative, counter-intuitive examples in low dimensions

Consider the following linear-quadratic optimization problem

$$\min_{x=(x_s,x_c)\in\mathbb{R}^n} \tfrac{1}{2}x^\mathsf{T}Hx, \quad \text{s.t. } (A_1 \, A_2)x = 0 \tag{7.5}$$

with symmetric positive-definite Hessian H and invertible A_1. The unique solution is $x^* = 0$. As before we approximate A_1 with $\widetilde{A}_1$ such that we obtain a contracting method for the forward problem. Without loss of generality, let $\widetilde{A}_1 = \mathbb{I}$ (otherwise multiply the constraint in (7.5) with $\widetilde{A}_1^{-1}$ from the left). We shall now have a look at instances of problem (7.5) with $n = m = 2$. We stress that there is nothing obviously pathologic about the following examples. The exact and approximated constraint Jacobians have full rank, the Hessians are symmetric positive-definite, and A_1 is always diagonalizable or even symmetric. We use the notation

$$A = \begin{pmatrix} A_1 & A_2 \end{pmatrix}, \quad \widetilde{A} = \begin{pmatrix} \widetilde{A}_1 & A_2 \end{pmatrix}, \quad K = \begin{pmatrix} H & A^\mathsf{T} \\ A & 0 \end{pmatrix}, \quad \widetilde{K} = \begin{pmatrix} H & \widetilde{A}^\mathsf{T} \\ \widetilde{A} & 0 \end{pmatrix}.$$

In all examples, the condition numbers of K and $\widetilde{K}$ are below 600.

7.1.1 Fast forward convergence, optimization divergence

As a first instance we investigate problem (7.5) for the special choice of

$$
\left(H \quad A^{\mathrm{T}} \right) = \left(
\begin{array}{cccc|cc}
0.67 & 0.69 & -0.86 & -0.13 & 1 & -0.072 \\
0.69 & 19 & 2.1 & -1.6 & -0.072 & 0.99 \\
-0.86 & 2.1 & 1.8 & -0.33 & -0.95 & 0.26 \\
-0.13 & -1.6 & -0.33 & 0.78 & -1.1 & -0.19
\end{array}
\right) , \quad (\text{Ex1})
$$

According to the Local Contraction Theorem 5.5 and Remark 5.6 the choice of $\widetilde{A}_1 = \mathbb{I}$ leads to a fast linear contraction rate for the forward problem of

$$
\kappa_F = \sigma_r(\mathbb{I} - \widetilde{A}_1^{-1} A_1) = \sigma_r(\mathbb{I} - A_1) \approx 0.077 < 1.
$$

However, for the contraction rate of the inexact SQP method with exact Hessian and exact constraint derivative with respect to x_c, we get

$$
\kappa_O = \sigma_r(\mathbb{I} - \widetilde{K}^{-1} K) \approx 1.07 > 1.
$$

Thus the full-step inexact SQP method does not have the property of linear local convergence. In fact it diverges if the starting point z^0 has a non-vanishing component in the direction of any generalized eigenvector of $\mathbb{I} - \widetilde{K}^{-1} K$ corresponding to a Jordan block with diagonal entries greater than 1.

7.1.2 Forward divergence, fast optimization convergence

Our second example is

$$
\left(H \quad A^{\mathrm{T}} \right) = \left(
\begin{array}{cccc|cc}
17 & 13 & 1.5 & -0.59 & 0.27 & -0.6 \\
13 & 63 & 7.3 & -4.9 & -0.6 & 0.56 \\
1.5 & 7.3 & 1.2 & -0.74 & -0.73 & -3.5 \\
-0.59 & -4.9 & -0.74 & 0.5 & -1.4 & -0.0032
\end{array}
\right) . \quad (\text{Ex2})
$$

We obtain

$$
\kappa_F = \sigma_r(\mathbb{I} - \widetilde{A}_1^{-1} A_1) \approx 1.20 > 1, \qquad \kappa_O = \sigma_r(\mathbb{I} - \widetilde{K}^{-1} K) \approx 0.014 < 1,
$$

i.e., fast convergence of the method for the optimization problem but divergence of the method for the forward problem. From these two examples we see that in general only little can be said about the connection between contraction for the forward and the optimization problem.

7.2 Subproblem regularization without changing the Jacobian approximation

We consider another example which exhibits de facto loss of convergence for Griewank's One-Step One-Shot method and for certain subproblem regularizations. By de facto loss of convergence we mean that although κ_F is well below 1 (e.g., below 0.5), κ_O is greater than 0.99. With

$$
\left(H \quad A^T\right) = \left(\begin{array}{cccc|cc}
0.83 & 0.083 & 0.34 & -0.21 & 1.1 & 0 \\
0.083 & 0.4 & -0.34 & -0.4 & 1.7 & 0.52 \\
0.34 & -0.34 & 0.65 & 0.48 & -0.55 & -1.4 \\
-0.21 & -0.4 & 0.48 & 0.75 & -0.99 & -1.8
\end{array}\right) \quad \text{(Ex3)}
$$

we obtain

$$
\kappa_F = \sigma_r(\mathbb{I} - \tilde{A}_1^{-1} A_1) \approx 0.48 < 1, \qquad \kappa_O = \sigma_r(\mathbb{I} - \tilde{K}^{-1} K) \approx 1.54 > 1.
$$

The quantities N_{xx}, G_y, G_u in the notation of Griewank [71] are

$$
N_{xx} = \mu H, \quad G_y = \mathbb{I} - \tilde{A}_1^{-1} A_1, \quad G_u = -A_2,
$$

where $\mu > 0$ is some chosen weighting factor for relative scaling of primal and dual variables. Based on

$$
Z(\lambda) = (\lambda \mathbb{I} - G_y)^{-1} G_u, \quad H(\lambda) = (Z(\lambda)^T, \mathbb{I}) N_{xx} (Z(\lambda)^T, \mathbb{I})^T,
$$

we can numerically verify that the projected Hessian preconditioners $H(\lambda)$, $\lambda \in [-1, 1]$, do not restore contraction. The lowest spectral radius of the iteration matrix is 1.17 for $\lambda = -0.57$ and larger for all other values (compare Figure 7.1).

We now investigate three different modifications of the subproblems which do not alter the Jacobian blocks of the KKT systems. These modifications are based on

$$
\kappa_O = \sigma_r(\mathbb{I} - \tilde{K}^{-1} K) = \sigma_r(\tilde{K}^{-1}(\tilde{K} - K)),
$$

which suggests that small eigenvalues of $\tilde{K}$ might lead to large κ_O. Thus we regularize $\tilde{K}$ such that the inverse $\tilde{K}^{-1}$ does not have large eigenvalues in the directions of inexactness of $\Delta K = \tilde{K} - K$.

We consider three prototypical regularization methods here which all add a positive multiple α of a matrix Λ to $\tilde{K}$. The regularizing matrices are

$$
\Lambda_p = \begin{pmatrix} \mathbb{I} & 0 & 0 \\ 0 & \mathbb{I} & 0 \\ 0 & 0 & 0 \end{pmatrix}, \quad \Lambda_{pd} = \begin{pmatrix} \mathbb{I} & 0 & 0 \\ 0 & \mathbb{I} & 0 \\ 0 & 0 & -\mathbb{I} \end{pmatrix}, \quad \Lambda_{hp} = \begin{pmatrix} 0 & 0 & 0 \\ 0 & \mathbb{I} & 0 \\ 0 & 0 & 0 \end{pmatrix},
$$

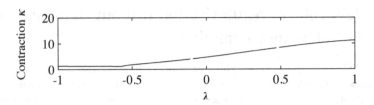

Figure 7.1: Contraction rates for One-Step One-Shot preconditioning with the projected Hessians $H(\lambda)$. The two gaps are due to the two eigenvalues of G_y rendering $\lambda \mathbb{I} - G_y$ singular.

where the subscripts stand for primal, primal-dual, and hemi-primal (i.e., only in the space of design variables), respectively.

7.3 Analysis of the regularized subproblems

We investigate the asymptotic behavior of the subproblem solution for $\alpha \to \infty$ for the primal, primal-dual, and hemi-primal regularization. We assume invertibility of the approximation $\widetilde{A}_{1k}$ and drop the iteration index k. We generally assume that H is positive-definite on the nullspace of the approximation $\widetilde{A}$.

Consider the α-dependent linear system for the step determination of the inexact SQP method

$$\left(\widetilde{K} + \alpha \Lambda \right) \begin{pmatrix} \Delta x(\alpha) \\ \Delta y(\alpha) \end{pmatrix} = \begin{pmatrix} -\ell \\ -r \end{pmatrix}, \tag{7.6}$$

where ℓ is the current Lagrange gradient and r is the current residual of the equality constraint. We use a nullspace method to solve the α-dependent system (7.6). Let matrices $Y \in \mathbb{R}^{n \times m}$ and $Z \in \mathbb{R}^{n \times (n-m)}$ have the properties

$$\widetilde{A} Z = 0, \quad (Z\,Y)^{\mathrm{T}} (Z\,Y) = \begin{pmatrix} Z^{\mathrm{T}} Z & 0 \\ 0 & Y^{\mathrm{T}} Y \end{pmatrix}, \quad \det(Y\,Z) \neq 0.$$

In other words, the columns of Z span the nullspace of $\widetilde{A}$. These are completed to form a basis of $\mathbb{R}^n$ by the columns of Y which are orthogonal to the columns of Z. In the new basis, we have $\Delta x = Y p + Z q$, with $(p, q) \in \mathbb{R}^n$.

7.3.1 Primal regularization

The motivation for the primal regularization stems from an analogy to the Levenberg-Marquardt method which, in the case of unconstrained minimization, is equivalent to a trust-region modification of the subproblem (see, e.g., Nocedal and Wright [121]). It turns out that the regularization with Λ_p bends the primal subproblem solutions towards the step of smallest Euclidean norm onto the linearized feasible set. However, it leads to a blow-up in the dual solution. From the following Lemma we observe that the primal step for large α is close to the step obtained by the Moore-Penrose-Pseudoinverse $\widetilde{A}^+ = \widetilde{A}^T(\widetilde{A}\widetilde{A}^T)^{-1}$ for the underdetermined system (7.1) and that the step in the Lagrange multiplier blows up for $r \neq 0$, and thus convergence cannot be expected.

Lemma 7.1. *Under the general assumptions of Section 7.3 the solution of equation (7.6) for the primal regularization for large α is asymptotically given by*

$$\Delta x(\alpha) = -\widetilde{A}^+ r + (1/\alpha)ZZ^+\left(H\widetilde{A}^+ r - \ell\right) + o(1/\alpha),$$
$$\Delta y(\alpha) = \alpha(\widetilde{A}\widetilde{A}^T)^{-1}r + (\widetilde{A}^+)^T(H\widetilde{A}^+ r - \ell) + o(1).$$

Proof. From the second block-row of equation (7.6) and the fact that $\widetilde{A}Y$ is invertible due to $\widetilde{A}$ having full rank we obtain $p = -(\widetilde{A}Y)^{-1}r$. Premultiplying the first block-row of equation (7.6) with Z^T from the left yields the α-dependent equation

$$Z^T HYp + Z^T HZq + \alpha Z^T Zq + Z^T \ell = 0. \tag{7.7}$$

Let $\alpha > 0$ and $\beta = 1/\alpha$. Solutions of equation (7.7) satisfy

$$\Phi(q,\beta) := \left(\beta Z^T HZ + Z^T Z\right)q + \beta Z^T\left(\ell - HY(\widetilde{A}Y)^{-1}r\right) = 0.$$

It holds that $\Phi(0,0) = 0$ and $\frac{\partial\Phi}{\partial q}(0,0) = Z^T Z$ is invertible, as Z has full rank. Therefore the Implicit Function Theorem yields the existence of a neighborhood $U \subset \mathbb{R}$ of 0 and a continuously differentiable function $\bar{q} : U \to \mathbb{R}^m$ such that $\bar{q}(0) = 0$ and

$$\Psi(\beta) := \Phi(\bar{q}(\beta),\beta) = 0 \quad \forall \beta \in U.$$

Using $0 = \frac{d\Psi}{d\beta} = \frac{\partial\Phi}{\partial q}\frac{d\bar{q}}{d\beta} + \frac{\partial\Phi}{\partial\beta}$ and Taylor's Theorem we have

$$\bar{q}(\beta) = \bar{q}(0) + \frac{d\bar{q}}{d\beta}(0)\beta + o(\beta) = \beta(Z^T Z)^{-1}Z^T\left(HY(\widetilde{A}Y)^{-1}r - \ell\right) + o(\beta),$$

which lends itself to the asymptotic

$$\Delta x(\alpha) = -Y(\widetilde{A}Y)^{-1}r + (1/\alpha)Z(Z^\mathsf{T}Z)^{-1}Z^\mathsf{T}\left(HY(\widetilde{A}Y)^{-1}r - \ell\right) + o(1/\alpha) \quad (7.8)$$

of the primal solution of equation (7.6) for large regularization parameters α.

Consider a special choice for the matrices Y and Z based on the QR decomposition $\widetilde{A} = Q\begin{pmatrix} R & B \end{pmatrix}$ with unitary Q and invertible R. We define

$$Z = \begin{pmatrix} -R^{-1}B \\ \mathbb{I} \end{pmatrix}, \qquad\qquad Y = \begin{pmatrix} R^\mathsf{T} \\ B^\mathsf{T} \end{pmatrix} = \widetilde{A}^\mathsf{T}Q$$

and obtain $Y(\widetilde{A}Y)^{-1} = \widetilde{A}^\mathsf{T}QQ^{-1}(\widetilde{A}\widetilde{A}^\mathsf{T})^{-1} = \widetilde{A}^+$, which yields the first assertion of the Lemma.

For the corresponding dual solution we multiply the first block-row of equation (7.6) with Y^T from the left to obtain

$$Y^\mathsf{T}(H + \alpha\mathbb{I})\Delta x(\alpha) + (\widetilde{A}Y)^\mathsf{T}\Delta\lambda(\alpha) + Y^\mathsf{T}\ell = 0.$$

After some rearrangements and with the help of the identity

$$(\widetilde{A}^+)^\mathsf{T}(\mathbb{I} - ZZ^+) = (\widetilde{A}Y)^{-\mathsf{T}}Y^\mathsf{T}(\mathbb{I} - Z(Z^\mathsf{T}Z)^{-1}Z^\mathsf{T}) = (\widetilde{A}^+)^\mathsf{T}$$

we obtain the second assertion of the Lemma. $\square$

7.3.2 Primal-dual regularization

The primal-dual regularization is motivated by moving all eigenvalues of the regularized KKT matrix away from zero. It is well known that under our assumptions the matrix $\widetilde{K}$ has $n + m$ positive and n negative eigenvalues (see Gould [63]). The primal regularization method only moves the positive eigenvalues away from zero. By adding the $-\mathbb{I}$ term to the lower right block, also the negative eigenvalues can be moved away from zero while conserving the inertia of $\widetilde{K}$.

Lemma 7.2. *Under the general assumptions of Section 7.3 the solution of equation (7.6) for the primal-dual regularization with large α is asymptotically given by*

$$\begin{pmatrix} \Delta x(\alpha) \\ \Delta y(\alpha) \end{pmatrix} = -\frac{1}{\alpha}\Lambda_{\mathrm{pd}}\begin{pmatrix} \ell \\ r \end{pmatrix} + o(1/\alpha) = \frac{1}{\alpha}\begin{pmatrix} -\ell \\ r \end{pmatrix} + o(1/\alpha).$$

Proof. Define again $\beta = 1/\alpha$, $z = (\Delta x, \Delta y)$, and

$$\Phi(z,\beta) = (\beta \widetilde{K} + \Lambda_{\text{pd}})z + \beta \begin{pmatrix} \ell \\ r \end{pmatrix}.$$

It holds that

$$\Phi(0,0) = 0, \qquad \frac{\partial \Phi}{\partial z} = \beta \widetilde{K} + \Lambda_{\text{pd}}, \qquad \frac{\partial \Phi}{\partial z}(0,0) = \Lambda_{\text{pd}}.$$

The Implicit Function Theorem and Taylor's Theorem yield the assertion. $\square$

Consider the limit case

$$\begin{pmatrix} \Delta x(\alpha) \\ \Delta y(\alpha) \end{pmatrix} = -\frac{1}{\alpha} \Lambda_{\text{pd}} \begin{pmatrix} \ell \\ r \end{pmatrix}$$

and the corresponding local contraction rate $\kappa_{\text{pd}} = \sigma_{\text{r}}(\mathbb{I} - (1/\alpha)\Lambda_{\text{pd}}\widetilde{K})$. If all the real parts of the (potentially complex) eigenvalues of the matrix $\Lambda_{\text{pd}}\widetilde{K}$ are larger than 0, contraction for large α can be recovered although contraction may be extremely slow, leading to de facto loss of convergence.

7.3.3 Hemi-primal regularization

In this section we are interested in a regularization of $\widetilde{K}$ only on the design variables x_{c} with Λ_{hp}. From the following Lemma we observe that for large α, the primal solution of the hemi-primal regularized subproblem tends toward the step obtained from equation (7.2) for the underdetermined system (7.1) and that the dual variables do not blow up for large α in the hemi-primal regularization.

Lemma 7.3. *Under the general assumptions of Section 7.3 the solution of equation (7.6) for the hemi-primal regularization is for large α asymptotically given by*

$$\Delta x(\alpha) = \begin{pmatrix} -\widetilde{A}_1^{-1} r \\ 0 \end{pmatrix} + (1/\alpha) Z Z^{\text{T}} \left(H \begin{pmatrix} \widetilde{A}_1^{-1} r \\ 0 \end{pmatrix} - \ell \right) + o(1/\alpha), \qquad (7.9\text{a})$$

$$\Delta y(\alpha) = \begin{pmatrix} \widetilde{A}_1^{-1} \\ 0 \end{pmatrix}^{\text{T}} \left(H \begin{pmatrix} \widetilde{A}_1^{-1} r \\ 0 \end{pmatrix} - \ell \right) + o(1), \qquad (7.9\text{b})$$

with the choice $Z = \left((-\widetilde{A}_1^{-1} A_2)^{\text{T}} \ \mathbb{I} \right)^{\text{T}}$ *and* $Y = (\widetilde{A}_1 \ A_2)^{\text{T}} = \widetilde{A}^{\text{T}}$.

Proof. By our general assumption $\widetilde{A}_1$ is invertible and the previous assumptions on Y and Z are satisfied. Again it holds that $Y(\widetilde{A}Y)^{-1} = \widetilde{A}^+$. We recover p as before. Let $\beta = 1/\alpha$. We can define an implicit function to determine $q(\beta)$ asymptotically via

$$\Phi(q,\beta) = (\beta Z^{\mathrm{T}} H Z + \mathbb{I})q + Y_2 p + \beta Z^{\mathrm{T}}(HYp + \ell),$$

where we used that $Z_2^{\mathrm{T}} Z_2 = \mathbb{I}$. It holds that $\Phi(-A_2^{\mathrm{T}}p, 0) = 0$ and $\frac{\partial \Phi}{\partial q}(-A_2^{\mathrm{T}}p, 0) = \mathbb{I}$. Thus the Implicit Function Theorem together with Taylor's Theorem yields

$$q(\beta) = -A_2^{\mathrm{T}}p - \beta Z^{\mathrm{T}}\left(HYp + \ell - HZA_2^{\mathrm{T}}p\right) + o(\beta).$$

By resubstitution of p and $q(1/\alpha)$ by the use of the identity

$$(Y - ZA_2^{\mathrm{T}})(\widetilde{A}Y)^{-1} = (\widetilde{A}_1^{-\mathrm{T}}\ 0)^{\mathrm{T}}$$

we recover the first assertion of the Lemma.

For the dual solution, we again multiply the first block-row of equation (7.6) with Y^{T} from the left to obtain

$$Y^{\mathrm{T}}\left(H + \alpha \begin{pmatrix} 0 & 0 \\ 0 & \mathbb{I} \end{pmatrix}\right)\Delta x(\alpha) + (\widetilde{A}Y)^{\mathrm{T}}\Delta y(\alpha) + Y^{\mathrm{T}}\ell = 0,$$

which after some rearrangements yields the second assertion. $\square$

Consider the limit case $\alpha \to \infty$. We recover from equation (7.9a) that

$$\Delta z_k = \begin{pmatrix} -\widetilde{A}_1^{-1} r_k \\ 0 \end{pmatrix}.$$

Hence $x_c^k = x_c^*$ stays constant and x_k converges to a feasible point x^* with the contraction rate κ_F of the Newton-type method for problem (7.1). For the asymptotic step in the dual variables we then obtain

$$\Delta y_k = -\widetilde{A}_1^{-\mathrm{T}}\left(\nabla_x f(x_k) + \nabla_x g(x_k) y_k\right) + \left(\widetilde{A}_1^{-\mathrm{T}}\ 0\right) H \left(\widetilde{A}_1^{-\mathrm{T}}\ 0\right)^{\mathrm{T}} r_k.$$

For the convergence of the coupled system with x_k and y_k let us consider the Jacobian of the iteration $(x_{k+1}, y_{k+1}) = T(x_k, y_k)$ (with suitably defined T)

$$\frac{dT}{d(x,y)} = \begin{pmatrix} \mathbb{I} - \widetilde{A}_1^{-1}\nabla_x g(x)^{\mathrm{T}} & 0 \\ * & \mathbb{I} - \widetilde{A}_1^{-\mathrm{T}}\nabla_x g(x) \end{pmatrix}.$$

Hence (x_k, y_k) converges with linear convergence rate κ_F, and y_k converges to

$$y^* = -\widetilde{A}_1^{-\mathrm{T}}\nabla_x f(x^*, y^*).$$

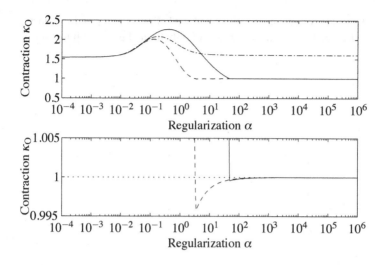

Figure 7.2: Divergence of the primal regularization $(-\cdot)$ and de facto loss of convergence for primal-dual $(--)$ and hemi-primal $(-)$ regularization for example Ex3 depending on the regularization value α. The lower diagram is a vertical close-up around $\kappa_O = 1$ of the upper diagram.

Thus the primal-dual iterates converge to a point which is feasible and stationary with respect to x_s but not necessarily to x_c. Taking α large but finite we see that the hemi-primal regularization damps design updates while correcting state and dual variables with a contraction of almost κ_F.

7.3.4 Divergence and de facto loss of convergence for subproblem regularizations

Figure 7.2 depicts the dependence of κ_O of the optimization method on the regularization parameter α and the choice of regularization (primal, primal-dual, and hemi-primal) on example Ex3. The example was specifically constructed to show de facto loss of convergence for all three regularizations. Obviously the primal regularization does not even reach $\kappa_O = 1$.

We want to remark that the above discussion is not a proof of convergence for the primal-dual or hemi-primal regularization approach. Nonetheless we have given a counter-example which shows failure of the primal regularization approach. With the de facto loss of convergence in mind we believe that a proof of convergence for the other regularization strategies is of only limited practical importance.

8 Condensing

Especially on fine space discretizations we obtain large scale quadratic subproblems (5.34) in the inexact SQP method described in Chapter 5. The goal of this chapter is to present a *condensing* approach which is one of two steps for the solution of these large scale QPs. It consists of a structure exploiting elimination of all discretized PDE variables from the QP. The resulting equivalent QP is of much smaller, grid-independent size and can then, in a second step, be solved by, e.g., a Parametric Active Set Method (PASM) which we describe in Chapter 9.

In Section 8.1 of this chapter we describe the typical multiple shooting structure. We highlight the additional Newton-Picard structures in Section 8.2. Then we present the exploitation of these structures for the elimination of the discretized PDE states in a rather general way in Section 8.3 and develop a particular Newton-Picard Hessian approximation which fits well in the condensing framework in Section 8.4. Based on the introduced notation we end this chapter with a result of scaling invariance of the Newton-Picard LISA-Newton method in Section 8.5.

8.1 Multiple shooting structure

In their seminal paper, Bock and Plitt [25] have described a condensing technique for the quadratic subproblems arising in an SQP method for Direct Multiple Shooting. We specialize this approach for the case that either fixed initial values or boundary value equality constraints are posed on the PDE states. In Section 8.2 we extend the approach with the exploitation of the Newton-Picard structure in the approximated Hessian and Jacobian matrices (see Chapter 6).

Recall the discretized NLP (3.3) on level l. To avoid notational clutter we drop the discretization level index l. The NLP (3.3) then reads

$$\underset{(q^i,s^i,v^i)_{i=0}^{n_{MS}}}{\text{minimize}} \; \Phi(s^{n_{MS}}, v^{n_{MS}}) \tag{8.1a}$$

$$\text{s.t.} \quad r_s^b(s^{n_{MS}}, v^{n_{MS}}) - s^0 = 0, \tag{8.1b}$$

$$r_v^b(s^{n_{MS}}, v^{n_{MS}}) - v^0 = 0, \tag{8.1c}$$

$$\bar{u}^i(t^i; q^{i-1}, s^{i-1}, v^{i-1}) - s^i = 0, \quad i = 1, \dots, n_{MS}, \tag{8.1d}$$

$$\bar{v}^i(t^i; q^{i-1}, s^{i-1}, v^{i-1}) - v^i = 0, \quad i = 1, \dots, n_{MS}, \tag{8.1e}$$

$$q^{n_{MS}} - q^{n_{MS}-1} = 0, \tag{8.1f}$$

$$r^i(q^{i-1}, v^{i-1}) \geq 0, \qquad i = 1, \dots, n_{MS}, \tag{8.1g}$$

$$r^e(v^{n_{MS}}) \geq 0, \tag{8.1h}$$

where we have split up the boundary condition r^b into two parts r_s^b and r_v^b. We abbreviate derivatives that occur in the remainder of this chapter according to

$$R_{ss}^b := \frac{\partial r_s^b}{\partial s^{n_{MS}}}, \quad R_{sv}^b := \frac{\partial r_s^b}{\partial v^{n_{MS}}}, \quad R_{vs}^b := \frac{\partial r_v^b}{\partial s^{n_{MS}}}, \quad R_{vv}^b := \frac{\partial r_v^b}{\partial v^{n_{MS}}},$$

$$G_q^i := \frac{\partial \bar{u}^i}{\partial q^{i-1}}, \qquad G_s^i := \frac{\partial \bar{u}^i}{\partial s^{i-1}}, \qquad G_v^i := \frac{\partial \bar{u}^i}{\partial v^{i-1}},$$

$$H_q^i := \frac{\partial \bar{v}^i}{\partial q^{i-1}}, \qquad H_s^i := \frac{\partial \bar{v}^i}{\partial s^{i-1}}, \qquad H_v^i := \frac{\partial \bar{v}^i}{\partial v^{i-1}},$$

$$R_q^{i,i} := \frac{\partial r^i}{\partial q^{i-1}}, \qquad R_v^{i,i} := \frac{\partial r^i}{\partial v^{i-1}}, \qquad R^e := \frac{r^e}{\partial v^{n_{MS}}}.$$

Let the Lagrangian of NLP (8.1) be denoted by $\mathscr{L}$. It is well-known (see, e.g., Bock and Plitt [25]) that due to the at most linear coupling between variables corresponding to shooting nodes i and $i+1$ the Hessian matrix of the Lagrangian $\mathscr{L}$ has block diagonal form. Our goal is to eliminate all PDE state variables s^i and so we regroup the variables in the order

$$(s^0, \dots, s^{n_{MS}}, v^0, \dots, v^{n_{MS}}, q^0, \dots, q^{n_{MS}}).$$

The elimination is based on the s^0-dependent part of boundary condition (8.1b) and on the matching conditions (8.1d). Thus we also shuffle the constraint order such that they are the first $n_s(n_{\mathrm{MS}}+1)$ constraints. We are lead to consider NLP (8.1) in the ordering

$$\underset{(x_1,x_2)\in\mathbb{R}^{n_1+n_2}}{\text{minimize}} \quad f(x_1,x_2) \tag{8.2a}$$

$$\text{s.t.} \quad g_i(x_1,x_2) = 0, \quad i \in \mathscr{E}_1, \tag{8.2b}$$

$$g_i(x_1,x_2) = 0, \quad i \in \mathscr{E}_2, \tag{8.2c}$$

$$g_i(x_1,x_2) \geq 0, \quad i \in \mathscr{I}, \tag{8.2d}$$

where $|\mathscr{E}_1| = n_1$. In Section 8.2 we describe how to exploit $g_i, i \in \mathscr{E}_1$, for partial reduction on the QP level. Now x_1 contains only the discretized PDE state variables and $g_i, i \in \mathscr{E}_1$, comprises the boundary and matching conditions (8.1b) and (8.1d).

Then the compound derivative of the constraints has the form

$$C = \begin{pmatrix} C_{11} & C_{12} \\ C_{21} & C_{22} \\ C_{31} & C_{32} \end{pmatrix}$$

$$= \left(\begin{array}{ccc|ccc|ccc} -\mathbb{I} & & & R_{ss}^{\mathrm{b}} & & & R_{sv}^{\mathrm{b}} & & \\ G_s^1 & -\mathbb{I} & & G_v^1 & & & G_q^1 & & \\ & \ddots & \ddots & & \ddots & & & \ddots & \\ & & G_s^{n_{\mathrm{MS}}} & -\mathbb{I} & & G_v^{n_{\mathrm{MS}}} & & & G_q^{n_{\mathrm{MS}}} \\ \hline & & & R_{vs}^{\mathrm{b}} & -\mathbb{I} & & R_{vv}^{\mathrm{b}} & & \\ H_s^1 & & & H_v^1 & -\mathbb{I} & & H_q^1 & & \\ & \ddots & & & \ddots & \ddots & & \ddots & \\ & & H_s^{n_{\mathrm{MS}}} & & H_v^{n_{\mathrm{MS}}} & -\mathbb{I} & & H_q^{n_{\mathrm{MS}}} \\ & & & & & & \mathbb{I} & -\mathbb{I} \\ \hline & & & R_v^{\mathrm{i},1} & & & R_q^{\mathrm{i},1} & & \\ & & & & \ddots & & & \ddots & \\ & & & R_v^{\mathrm{i},n_{\mathrm{MS}}} & & & R_q^{\mathrm{i},n_{\mathrm{MS}}} & & \\ & & & R_v^{\mathrm{e},n_{\mathrm{MS}}} & & & & & \end{array} \right).$$

We want to stress that contrary to the appearance the block C_{11} is several orders of magnitude larger than the blocks C_{22} and C_{32} on fine spatial discretization levels.

The next lemma shows that under suitable assumptions C_{11} is invertible. We use products of non-commuting matrices where the order is defined via

$$\prod_{i=1}^{n_{\mathrm{MS}}} G_s^i := G_s^{n_{\mathrm{MS}}} \cdots G_s^1 \quad \text{and} \quad \prod_{i=1}^{0} G_s^i = \mathbb{I} \text{ by convention.}$$

Lemma 8.1. *Let* $M_B = \mathbb{I} - (\prod_{i=1}^{n_{MS}} G_s^i) R_{ss}^b$. *If* M_B *is invertible so is* C_{11} *and the inverse is given by* $C_{11}^{-1} =$

$$
-\begin{pmatrix}
\mathbb{I} & & \left(\displaystyle\prod_{i=1}^{0} G_s^i\right) R_{ss}^b \\
& \ddots & \vdots \\
& \mathbb{I} & \left(\displaystyle\prod_{i=1}^{n_{MS}-1} G_s^i\right) R_{ss}^b \\
& & \mathbb{I}
\end{pmatrix}
\begin{pmatrix}
\mathbb{I} & & \\
& \ddots & \\
& & \mathbb{I} \\
& & & M_B^{-1}
\end{pmatrix}
\begin{pmatrix}
\mathbb{I} & & & \\
\displaystyle\prod_{i=1}^{1} G_s^i & \mathbb{I} & & \\
\vdots & & \ddots & \ddots \\
\displaystyle\prod_{i=1}^{n_{MS}} G_s^i & \cdots & \displaystyle\prod_{i=n_{MS}}^{n_{MS}} G_s^i & \mathbb{I}
\end{pmatrix}.
$$

Proof. We premultiply C_{11} with the matrices in the assertion one after the other to obtain

$$
\begin{pmatrix}
\mathbb{I} & & \\
G_s^1 & \mathbb{I} & \\
\vdots & & \ddots & \ddots \\
\displaystyle\prod_{i=1}^{n_{MS}} G_s^i & \cdots & G_s^{n_{MS}} & \mathbb{I}
\end{pmatrix}
\begin{pmatrix}
-\mathbb{I} & & & R_{ss}^b \\
G_s^1 & -\mathbb{I} & & \\
& \ddots & \ddots & \\
& & G_s^{n_{MS}} & -\mathbb{I}
\end{pmatrix}
=
\begin{pmatrix}
-\mathbb{I} & & \left(\displaystyle\prod_{i=1}^{0} G_s^i\right) R_{ss}^b \\
& -\mathbb{I} & \left(\displaystyle\prod_{i=1}^{1} G_s^i\right) R_{ss}^b \\
& & \ddots & \vdots \\
& & & -M_B
\end{pmatrix},
$$

$$
\begin{pmatrix}
\mathbb{I} & & \\
& \ddots & \\
& & \mathbb{I} \\
& & & M_B^{-1}
\end{pmatrix}
\begin{pmatrix}
-\mathbb{I} & & \left(\displaystyle\prod_{i=1}^{0} G_s^i\right) R_{ss}^b \\
& -\mathbb{I} & \left(\displaystyle\prod_{i=1}^{1} G_s^i\right) R_{ss}^b \\
& & \ddots & \vdots \\
& & & -M_B
\end{pmatrix}
=
\begin{pmatrix}
-\mathbb{I} & & \left(\displaystyle\prod_{i=1}^{0} G_s^i\right) R_{ss}^b \\
& -\mathbb{I} & \left(\displaystyle\prod_{i=1}^{1} G_s^i\right) R_{ss}^b \\
& & \ddots & \vdots \\
& & & -\mathbb{I}
\end{pmatrix},
$$

$$
\begin{pmatrix}
-\mathbb{I} & & -\left(\displaystyle\prod_{i=1}^{0} G_s^i\right) R_{ss}^b \\
& \ddots & \vdots \\
& & -\mathbb{I} - \left(\displaystyle\prod_{i=1}^{n_{MS}-1} G_s^i\right) R_{ss}^b \\
& & -\mathbb{I}
\end{pmatrix}
\begin{pmatrix}
-\mathbb{I} & & \left(\displaystyle\prod_{i=1}^{0} G_s^i\right) R_{ss}^b \\
& -\mathbb{I} & \left(\displaystyle\prod_{i=1}^{1} G_s^i\right) R_{ss}^b \\
& & \ddots & \vdots \\
& & & -\mathbb{I}
\end{pmatrix}
= \mathbb{I}.
$$

This proves the assertion. $\Box$

The assumption of invertibility of M_B is merely that one is not an eigenvalue of the matrix

$$
G_B := \left(\prod_{i=1}^{n_{MS}} G_s^i\right) R_{ss}^b
$$

which coincides with the monodromy matrix of the periodicity condition (8.1b) in the solution of NLP (8.1).

8.2 Newton-Picard structure

Now we investigate the structures arising from the approximation of the blocks in C via Newton-Picard (see Chapter 6). We want to stress that it does not matter if the two-grid or classical version of Newton-Picard is applied. We only assume that there exist a prolongation operator P and a restriction operator R which satisfy

$$RP = \mathbb{I}. \tag{8.3}$$

We now approximate the blocks in C. Let hatted matrices ($\hat{\ }$) denote the evaluation of a matrix on either a coarse grid (two-grid variant) or on the dominant subspace (classical variant). Then we assemble the approximations ($\tilde{\ }$) from the hatted matrices preceded and/or succeeded by appropriate prolongation and/or restriction matrices according to

$$\widetilde{R}^{b}_{ss} = P\hat{R}^{b}_{ss}R, \qquad \widetilde{R}^{b}_{sv} = P\hat{R}^{b}_{sv}, \qquad \widetilde{R}^{b}_{vs} = \hat{R}^{b}_{vs}R, \qquad \widetilde{R}^{b}_{vv} = \hat{R}^{b}_{vv}, \tag{8.4a}$$

$$\widetilde{G}^{i}_{q} = P\hat{G}^{i}_{q}, \qquad \widetilde{G}^{i}_{s} = P\hat{G}^{i}_{s}R, \qquad \widetilde{G}^{i}_{v} = P\hat{G}^{i}_{s}, \tag{8.4b}$$

$$\widetilde{H}^{i}_{q} = \hat{H}^{i}_{q}, \qquad \widetilde{H}^{i}_{s} = \hat{H}^{i}_{s}R, \qquad \widetilde{H}^{i}_{v} = \hat{H}^{i}_{v}, \tag{8.4c}$$

$$\widetilde{R}^{i,i}_{q} = \hat{R}^{i,i}_{q}, \qquad \widetilde{R}^{i,i}_{v} = \hat{R}^{i,i}_{v}, \qquad \widetilde{R}^{e} = \hat{R}^{e}. \tag{8.4d}$$

The following lemma shows that the approximation of M_B can be cheaply evaluated and inverted because it only involves operations on the coarse grid or on the low-dimensional dominant subspace.

Lemma 8.2. *Let*

$$\hat{G}_B := \left(\prod_{i=1}^{n_{MS}} \hat{G}^{i}_{s}\right) \hat{R}^{b}_{ss}, \qquad \widetilde{G}_B := \left(\prod_{i=1}^{n_{MS}} \widetilde{G}^{i}_{s}\right) \widetilde{R}^{b}_{ss},$$

$$\hat{M}_B := \mathbb{I} - \hat{G}_B, \qquad \widetilde{M}_B := \mathbb{I} - \widetilde{G}_B.$$

If $\hat{M}_B$ is invertible so is $\widetilde{M}_B$ and it holds that

$$\widetilde{M}_B^{-1} = \left(\mathbb{I} - P\hat{G}_B R\right)^{-1} = \mathbb{I} - P\left(\mathbb{I} - \hat{M}_B^{-1}\right) R.$$

Proof. From equation (8.3) we obtain

$$\widetilde{G}_B = \left(\prod_{i=1}^{n_{MS}} P\hat{G}^{i}_{s}R\right) P\hat{R}^{b}_{ss}R = P\hat{G}_B R$$

and thus

$$\widetilde{M}_B = \mathbb{I} - P\hat{G}_B R.$$

The calculation

$$\widetilde{M}_B \widetilde{M}_B^{-1} = \left(\mathbb{I} - P \hat{G}_B R \right) \left(\mathbb{I} - P \left[\mathbb{I} - \left(\mathbb{I} - \hat{G}_B \right)^{-1} \right] R \right)$$
$$= \mathbb{I} - P \hat{G}_B R - P \left(\mathbb{I} - \hat{G}_B \right) \left[\mathbb{I} - \left(\mathbb{I} - \hat{G}_B \right)^{-1} \right] R$$
$$= \mathbb{I} - P \left[\hat{G}_B + \mathbb{I} - \hat{G}_B - \mathbb{I} \right] R = \mathbb{I}$$

yields the assertion. $\square$

The structure of C and the assertion of Lemma 8.1 is also preserved if we use the proposed approximations. Thus Lemma 8.2 suggests that it is possible to compute the inverse of the approximation of the large block C_{11} in a cheap way. We prove this supposition in Theorem 8.3 but before we need to introduce another notational convention, the Kronecker product of two matrices. We only use the special case where the left-hand factor is the identity and thus we have for an arbitrary matrix A that

$$\mathbb{I}_{p \times p} \otimes A := \begin{pmatrix} A & & \\ & \ddots & \\ & & A \end{pmatrix} \quad (p \text{ instances of } A \text{ blocks on the diagonal}).$$

Theorem 8.3. *Define the projectors*

$$\Pi^{\text{slow}} = \mathbb{I}_{n_{MS} \times n_{MS}} \otimes (PR), \quad \Pi^{\text{fast}} = \mathbb{I} - \Pi^{\text{slow}}.$$

Then

$$\widetilde{C}_{11}^{-1} \Pi^{\text{slow}} = (\mathbb{I} \otimes P) \hat{C}_{11}^{-1} (\mathbb{I} \otimes R), \quad \widetilde{C}_{11}^{-1} \Pi^{\text{fast}} = -\Pi^{\text{fast}}.$$

Proof. Lemma 8.1 yields a decomposition of C_{11}^{-1} into a product of the three matrices $A_1 A_2 A_3$. The same type of decomposition holds when the blocks in C are substituted by their tilde or hat counterparts. We now show in three steps that

$$\widetilde{A}_k (\mathbb{I} \otimes P) = (\mathbb{I} \otimes P) \hat{A}_k, \quad k = 1, 2, 3,$$

from which we can immediately infer the assertion

$$\widetilde{C}_{11}^{-1} \Pi^{\text{slow}} = \widetilde{A}_1 \widetilde{A}_2 \widetilde{A}_3 (\mathbb{I} \otimes (PR)) = (\mathbb{I} \otimes P) \hat{A}_1 \hat{A}_2 \hat{A}_3 (\mathbb{I} \otimes R) = (\mathbb{I} \otimes P) \hat{C}_{11}^{-1} (\mathbb{I} \otimes R).$$

The cases $k = 1, 3$ follow from

$$\left(\prod_{i=j_1}^{j_2} \widetilde{G}_s^i \right) P = \left(\prod_{i=j_1}^{j_2} (P \hat{G}_s^i R) \right) P = P \left(\prod_{i=j_1}^{j_2} \hat{G}_s^i \right),$$
$$\widetilde{R}_{ss}^{\text{b}} P = P \hat{R}_{ss}^{\text{b}} R P = P \hat{R}_{ss}^{\text{b}}.$$

The case $k = 2$ only involves the inverse given by Lemma 8.2

$$\widetilde{M}_B^{-1} P = \left(\mathbb{I} - P \left(\mathbb{I} - \hat{M}_B^{-1} \right) R \right) P = P \hat{M}_B^{-1}.$$

The proof of the assertion for Π^{fast} is based on equation (8.3) which yields

$$R(\mathbb{I} - PR) = R - RPR = 0. \tag{8.5}$$

We obtain

$$\widetilde{A}_1 \Pi^{\text{fast}} = -\Pi^{\text{fast}}, \qquad \widetilde{A}_2 \Pi^{\text{fast}} = \Pi^{\text{fast}}, \qquad \widetilde{A}_3 \Pi^{\text{fast}} = \Pi^{\text{fast}},$$

because all off-diagonal blocks are eliminated due to equation (8.5) and

$$\widetilde{M}_B^{-1}(\mathbb{I} - PR) = \left(\mathbb{I} - P \left(\mathbb{I} - \hat{M}_B^{-1} \right) R \right) (\mathbb{I} - PR) = \mathbb{I} - PR.$$

From equation (8.5) follows also immediately that

$$\Pi^{\text{fast}} \Pi^{\text{fast}} = \mathbb{I} \otimes \left[(\mathbb{I} - PR)(\mathbb{I} - PR) \right] = \Pi^{\text{fast}},$$

i.e., Π^{fast} is idempotent and thus indeed a projector. Hence we obtain

$$\widetilde{C}_{11}^{-1} \Pi^{\text{fast}} = \widetilde{A}_1 \widetilde{A}_2 \widetilde{A}_3 \Pi^{\text{fast}} = -\Pi^{\text{fast}}$$

which shows the last assertion. $\square$

Corollary 8.4. *If it exists, the Newton-Picard approximation of block C_{11} has the inverse*

$$\widetilde{C}_{11}^{-1} = (\mathbb{I} \otimes P) \left(\hat{C}_{11}^{-1} + \mathbb{I} \right) (\mathbb{I} \otimes R) - \mathbb{I}.$$

Proof. Consider

$$\widetilde{C}_{11}^{-1} = \widetilde{C}_{11}^{-1} \Pi^{\text{slow}} + \widetilde{C}_{11}^{-1} \Pi^{\text{fast}} = (\mathbb{I} \otimes P) \hat{C}_{11}^{-1} (\mathbb{I} \otimes R) + (\mathbb{I} \otimes (PR)) - \mathbb{I}$$
$$= (\mathbb{I} \otimes P) \left(\hat{C}_{11}^{-1} + \mathbb{I} \right) (\mathbb{I} \otimes R) - \mathbb{I},$$

which follows directly from Theorem 8.3. $\square$

Remark 8.5. The inversion of $\widetilde{C}_{11}$ via the formula from Corollary 8.4 also reduces the number of needed restrictions to the minimum of n_{MS}. This is important for FEM discretizations where an L^2 restriction involves the inversion of a reduced mass matrix.

Thus we see that the condensing operations involving $\widetilde{C}_{11}$ can be efficiently computed involving only operations on the coarse grid or on the small Newton-Picard subspace plus n_{MS} prolongations and restrictions.

8.3 Elimination of discretized PDE states

We now consider QPs with a structure inherited from NLP (8.2)

$$\underset{(x_1,x_2)\in\mathbb{R}^{n_1+n_2}}{\text{minimize}} \quad \frac{1}{2}\begin{pmatrix} x_1 \\ x_2 \end{pmatrix}^{\mathrm{T}}\begin{pmatrix} B_{11} & B_{12} \\ B_{21} & B_{22} \end{pmatrix}\begin{pmatrix} x_1 \\ x_2 \end{pmatrix} + \begin{pmatrix} b_1 \\ b_2 \end{pmatrix}^{\mathrm{T}}\begin{pmatrix} x_1 \\ x_2 \end{pmatrix} \tag{8.6a}$$

$$\text{s.t.} \quad C_{11}x_1 + C_{12}x_2 = c_1, \tag{8.6b}$$

$$C_{21}x_1 + C_{22}x_2 = c_2, \tag{8.6c}$$

$$C_{31}x_1 + C_{32}x_2 \geq c_3. \tag{8.6d}$$

Imagine the variable vector x_1 comprising all discretized PDE states and the small variable vector x_2 containing the remaining degrees of freedom. The proof of the following theorem can be carried out via KKT transformation rules as in Leineweber [106]. We want to give a slightly shorter proof here which can be interpreted as a partial null-space approach.

Theorem 8.6. *Assume that C_{11} in QP (8.6) is invertible and define*

$$Z = \begin{pmatrix} -C_{11}^{-1}C_{12} \\ \mathbb{I} \end{pmatrix}, \qquad B' = Z^{\mathrm{T}}BZ,$$

$$c_1' = C_{11}^{-1}c_1, \qquad b' = B_{21}c_1' + b_2 - C_{12}^{\mathrm{T}}C_{11}^{-\mathrm{T}}(B_{11}c_1' + b_1),$$

$$c_2' = c_2 - C_{21}c_1', \qquad C_2' = C_{22} - C_{21}C_{11}^{-1}C_{12},$$

$$c_3' = c_3 - C_{31}c_1', \qquad C_3' = C_{32} - C_{31}C_{11}^{-1}C_{12}.$$

Let furthermore $(x_2^*, y_{\mathscr{E}_2}^*, y_{\mathscr{I}}^*) \in \mathbb{R}^{n_2+m_2+m_3}$ be a primal-dual solution of the QP

$$\underset{x_2\in\mathbb{R}^{n_2}}{\text{minimize}} \quad \frac{1}{2}x_2^{\mathrm{T}}B'x_2 + b'^{\mathrm{T}}x_2 \quad \text{s.t.} \quad C_2'x_2 = c_2', \quad C_3'x_2 \geq c_3'. \tag{8.7}$$

If we choose

$$x_1^* = C_{11}^{-1}(c_1 - C_{12}x_2^*), \tag{8.8a}$$

$$y_{\mathscr{E}_1}^* = C_{11}^{-\mathrm{T}}\left((B_{12} - B_{11}C_{11}^{-1}C_{12})x_2^* + B_{11}c_1' + b_1 - C_{21}^{\mathrm{T}}y_{\mathscr{E}_2}^* - C_{31}^{\mathrm{T}}y_{\mathscr{I}}^*\right) \tag{8.8b}$$

then $(x^*, y^*) := (x_1^*, x_2^*, y_{\mathscr{E}_1}^*, y_{\mathscr{E}_2}^*, y_{\mathscr{I}}^*)$ is a primal-dual solution of QP (8.6).

Proof. We first observe that constraint (8.6b) is equivalent to equation (8.8a) and that thus

$$\begin{pmatrix} x_1^* \\ x_2^* \end{pmatrix} = Zx_2^* + \begin{pmatrix} c_1' \\ 0 \end{pmatrix}$$

is satisfied. Let us define

$$Y = \begin{pmatrix} \mathbb{I} \\ 0 \end{pmatrix}.$$

The unit upper triangular matrix $Q := \begin{pmatrix} Y & Z \end{pmatrix}$ is invertible and we can multiply the stationarity condition of QP (8.6) from the left with Q^T to obtain the equivalent system of equations

$$0 = Y^T B Z x_2^* + B_{11} c_1' + b_1 - C_{11}^T y_{\mathcal{E}_1}^* - C_{21}^T y_{\mathcal{E}_2}^* - C_{31}^T y_{\mathcal{I}}^*, \tag{8.9a}$$

$$0 = Z^T B Z x_2^* + Z^T \left(\begin{pmatrix} B_{11} \\ B_{21} \end{pmatrix} c_1' + b \right) - Z^T \begin{pmatrix} C_{11} & C_{12} \\ C_{21} & C_{22} \\ C_{31} & C_{32} \end{pmatrix}^T \begin{pmatrix} y_{\mathcal{E}_1}^* \\ y_{\mathcal{E}_2}^* \\ y_{\mathcal{I}}^* \end{pmatrix}. \tag{8.9b}$$

Expansion of $Y^T B Z$ yields that condition (8.9a) is equivalent to equation (8.8b) and by virtue of

$$CZ = \begin{pmatrix} C_{11} & C_{12} \\ C_{21} & C_{22} \\ C_{31} & C_{32} \end{pmatrix} \begin{pmatrix} -C_{11}^{-1} C_{12} \\ \mathbb{I} \end{pmatrix} = \begin{pmatrix} 0 \\ C_2' \\ C_3' \end{pmatrix}$$

condition (8.9b) is equivalent to the stationarity condition of QP (8.7)

$$B' x_2^* + b' - {C_2'}^T y_{\mathcal{E}_2}^* - {C_3'}^T y_{\mathcal{I}}^* = 0.$$

Feasibility is guaranteed by

$$C_{21} x_1^* + C_{22} x_2^* = C_{21} C_{11}^{-1} (c_1 - C_{12} x_2^*) + C_{22} x_2^* = c_2 \quad \Leftrightarrow \quad C_2' x_2^* = c_2',$$
$$C_{31} x_1^* + C_{32} x_2^* = C_{31} C_{11}^{-1} (c_1 - C_{12} x_2^*) + C_{32} x_2^* \geq c_3 \quad \Leftrightarrow \quad C_3' x_2^* \geq c_3'.$$

Finally complementarity holds because the multipliers $y_{\mathcal{I}}^*$ for the inequalities are the same in the condensed QP (8.7) and in the structured QP (8.6). $\square$

The condensed QP (8.7) is of much smaller size than QP (8.6) and its size does not depend on the spatial discretization level. It still exhibits the typical multiple shooting structure in the ODE states and could thus be condensed one more time. In the examples which we present in Part III, however, the computational savings are only marginal between skipping the second condensing and solving QP (8.7) directly with the method we describe in Chapter 9.

8.4 Newton-Picard Hessian approximation

We can efficiently evaluate the quantities that must be computed to set up the
condensed QP (8.7) of Theorem 8.6 by once again exploiting the Newton-Picard
structure in the approximations of the constraint Jacobian: The partial null-space
basis can be evaluated purely on the slow subspace because

$$
\widetilde{Z} = \begin{pmatrix} -\widetilde{C}_{11}^{-1}\widetilde{C}_{12} \\ \mathbb{I} \end{pmatrix} = \begin{pmatrix} -(\mathbb{I}\otimes P)\hat{C}_{11}^{-1}(\mathbb{I}\otimes R)(\mathbb{I}\otimes P)\hat{C}_{12} \\ \mathbb{I} \end{pmatrix}
$$

$$
= \begin{pmatrix} -(\mathbb{I}\otimes P)\hat{C}_{11}^{-1}\hat{C}_{12} \\ \mathbb{I} \end{pmatrix}.
$$

This observation suggests a projected Newton-Picard approximation of the Hessian matrix via

$$
\widetilde{B}^{\mathrm{fast}} = \begin{pmatrix} (\mathbb{I}\otimes \Pi^{\mathrm{fast}})^{\mathrm{T}}B_{11}(\mathbb{I}\otimes \Pi^{\mathrm{fast}}) & 0 \\ 0 & 0 \end{pmatrix},
$$

$$
\widetilde{B}^{\mathrm{slow}} = \begin{pmatrix} (\mathbb{I}\otimes R)^{\mathrm{T}}\hat{B}_{11}(\mathbb{I}\otimes R) & (\mathbb{I}\otimes R)^{\mathrm{T}}\hat{B}_{12} \\ \hat{B}_{21}(\mathbb{I}\otimes R) & \hat{B}_{22} \end{pmatrix},
$$

$$
\widetilde{B} = \widetilde{B}^{\mathrm{fast}} + \widetilde{B}^{\mathrm{slow}}.
$$

Consequently we have

$$
\widetilde{Z}^{\mathrm{T}}\widetilde{B}^{\mathrm{fast}}\widetilde{Z} = 0
$$

and thus we can also compute the condensed Newton-Picard Hessian matrix purely
on the slow subspace according to

$$
\widetilde{B}' = \widetilde{Z}^{\mathrm{T}}\widetilde{B}\widetilde{Z} = \hat{Z}^{\mathrm{T}}\hat{B}\hat{Z} \quad \text{with } \hat{Z} = \begin{pmatrix} -\hat{C}_{11}^{-1}\hat{C}_{12} \\ \mathbb{I} \end{pmatrix}.
$$

The Hessian term $\widetilde{B}^{\mathrm{fast}}$ only plays a role in the evaluation of $\widetilde{B}c_1'$. Thus we only
need to evaluate one matrix vector product with the exact Hessian matrix for the
solution of the large structured Newton-Picard quadratic subproblem. In the two-
grid variant all remaining matrix vector products with the approximated Hessian
only require the coarse-grid operations.

Numerical experience on the application problems of Part III has shown that
a pure coarse grid Hessian approximation leads to a substantial loss of contraction for the Newton-Picard LISA-Newton method while the contraction with the
Newton-Picard Hessian approximation yields contractions which are almost as
good as when a pure fine grid Hessian is used.

8.5 Scaling invariance of the Newton-Picard LISA-Newton method

Based on Corollary 5.25 we know that if a preconditioner respects the transformation property of Lemma 5.24 we obtain affine invariance of the LISA-Newton method. We now show that the Newton-Picard preconditioners partly satisfy the transformation property. As a result we obtain invariance of the Newton-Picard LISA-Newton method with respect to scaling. To be precise let $\alpha, \beta \in \mathbb{R}$, and

$$a^1 \in \mathbb{R}^{n_1}, \quad a_i^1 = \alpha, i \in 1,\ldots,n_1, \quad a^2 \in \mathbb{R}^{|\mathscr{E}_2|}, \quad a^3 \in \mathbb{R}^{|\mathscr{I}|}, \quad a = (a^1, a^2, a^3),$$

$$d^1 \in \mathbb{R}^{n_1}, \quad d_i^1 = \beta, i \in 1,\ldots,n_1, \quad d^2 \in \mathbb{R}_+^{n_2}, \qquad\qquad d = (d^1, d^2),$$

and assume that no entry of a and d vanishes. Moreover we define

$$A_i = \operatorname{diag}(a^i), i = 1,2,3, \quad A = \operatorname{diag}(a), \quad D_i = \operatorname{diag}(d^i), i = 1,2, \quad D = \operatorname{diag}(d).$$

We now consider the family of scaled NLPs of the form of NLP (8.2)

$$\underset{x=(x_1,x_2)\in\mathbb{R}^{n_1+n_2}}{\text{minimize}} \quad f(\beta x_1, D_2 x_2) \tag{8.10a}$$

$$\text{s.t.} \quad \alpha g_i(\beta x_1, D_2 x_2) = 0, \quad i \in \mathscr{E}_1, \tag{8.10b}$$

$$a_i g_i(\beta x_1, D_2 x_2) = 0, \quad i \in \mathscr{E}_2, \tag{8.10c}$$

$$a_i g_i(\beta x_1, D_2 x_2) \geq 0, \quad i \in \mathscr{I}. \tag{8.10d}$$

With $y = (y_1, y_2, y_3) \in \mathbb{R}^{n_1 + |\mathscr{E}_2| + |\mathscr{I}|}$ we obtain the scaled Lagrangian

$$\mathscr{L}^{\text{sc}}(x,y) = f(Dx) - \sum_{i\in\mathscr{E}_1} \alpha y_i g_i(Dx) - \sum_{i\in\mathscr{E}_2\cup\mathscr{I}} a_i y_i g_i(Dx)$$

and its gradient

$$\nabla_x \mathscr{L}^{\text{sc}}(x,y) = D\nabla f(Dx) - \sum_{i\in\mathscr{E}_1} \alpha y_i D\nabla g_i(Dx) - \sum_{i\in\mathscr{E}_2\cup\mathscr{I}} a_i y_i D\nabla g_i(Dx),$$

After introducing the scaled variables $x^{\text{sc}} = D^{-1}x$ and $y^{\text{sc}} = A^{-1}y$ we can establish for the function F of Section 5.7.1 in Chapter 5 and its scaled counterpart that

$$F^{\text{sc}}(x^{\text{sc}}, y^{\text{sc}}) = \begin{pmatrix} D\nabla_x\mathscr{L}(x,y) \\ Ag(Dx) \end{pmatrix} = \operatorname{diag}(D,A)F(x,y),$$

$$\frac{\mathrm{d}F^{\text{sc}}(x^{\text{sc}}, y^{\text{sc}})}{\mathrm{d}(x^{\text{sc}}, y^{\text{sc}})} = \begin{pmatrix} D & 0 \\ 0 & A \end{pmatrix} \frac{\mathrm{d}F(x,y)}{\mathrm{d}(x,y)} \begin{pmatrix} D & 0 \\ 0 & A \end{pmatrix}$$

According to Lemma 5.24 we need to verify that the Newton-Picard approximation satisfies the same transformation rule. Let $\hat{A}_1 = \text{diag}(\alpha)$ and $\hat{D}_1 = \text{diag}(\beta)$ denote the scaling matrices corresponding to A_1 and D_1 on the coarse grid. Then we can immediately see that the transformation rule for the blocks of C in equations (8.4) and for the two-grid Newton-Picard Hessian approximation of Section 8.4 satisfies the assumption for Lemma 5.24 due to

$$(\mathbb{I} \otimes P)\hat{A}_1 = \alpha \mathbb{I} \otimes P = A_1(\mathbb{I} \otimes P) \quad \text{and} \quad \hat{D}_1(\mathbb{I} \otimes R) = \beta \mathbb{I} \otimes R = (\mathbb{I} \otimes R)D_1.$$

Thus the Newton-Picard LISA-Newton method is scaling invariant.

9 A Parametric Active Set method for QP solution

Most of this chapter is an excerpt form the technical report Potschka et al. [129] which we include here for completeness. The chapter is dedicated to the numerical solution of the convex QP

$$\underset{x\in\mathbb{R}^n}{\text{minimize}}\, \frac{1}{2}x^{\mathrm{T}}Bx + b^{\mathrm{T}}x \quad \text{s.t.} \quad c^{\mathrm{l}} \leq Cx \leq c^{\mathrm{u}}, \tag{9.1}$$

with symmetric Hessian matrix $B \in \mathbb{R}^{n\times n}$, constraint matrix $C \in \mathbb{R}^{m\times n}$, gradient vector $b \in \mathbb{R}^n$, and lower and upper constraint vectors $c^{\mathrm{l}}, c^{\mathrm{u}} \in \mathbb{R}^m$. For most of this chapter we furthermore assume B to be positive semidefinite. We describe the generalization to nonconvex problems with indefinite B in Section 9.8.

The structure of this chapter is the following: We start with recalling the Parametric Quadratic Programming (PQP) method [17] in Section 9.2 and identify its fundamental numerical challenges in Section 9.3. In Section 9.4 we develop strategies to meet these challenges. It follows a short description of the newly developed Matlab® code rpasm in Section 9.5 which we compare with other popular academic and commercial QP solvers in Section 9.6. We continue in Section 9.7 with a description of drawbacks of the reliability improving strategies. In Section 9.8 we discuss an extension to compute local minima of nonconvex QPs.

9.1 General remarks on Quadratic Programming Problems

Although we are mainly concerned with QPs which arise as subproblems of the inexact SQP method described in Chapter 5, in particular after a condensing step according to Chapter 8, the class of convex QP problems is important in its own right. Gould and Toint [66] have been compiling a bibliography of currently 979 publications which comprises many application papers from disciplines as diverse as portfolio analysis, structural analysis, VLSI design, discrete-time stabilization, optimal and fuzzy control, finite impulse response design, optimal power flow, economic dispatch, etc. Several benchmark and application problems are collected in a repository [111] which is accessible through the CUTEr testing environment [68].

9.1.1 Optimality conditions

For the characterization of solutions of QP (9.1) we partition the index set $\overline{m} = \{1,\ldots,m\}$ into four disjoint sets

$$\mathscr{A}^e(x) = \{i \in \overline{m} \mid c_i^l = (Cx)_i = c_i^u\}, \quad \mathscr{A}^l(x) = \{i \in \overline{m} \mid c_i^l = (Cx)_i < c_i^u\},$$
$$\mathscr{A}^u(x) = \{i \in \overline{m} \mid c_i^l < (Cx)_i = c_i^u\}, \quad \mathscr{A}^f(x) = \{i \in \overline{m} \mid c_i^l < (Cx)_i < c_i^u\}$$

of equality, lower active, upper active, and free constraint indices, respectively. It is well known (see Chapter 4) that for any solution x^* of QP (9.1) there exists a vector $y^* \in \mathbb{R}^m$ of dual variables such that

$$Bx^* + b - C^T y^* = 0, \qquad\qquad c^l \leq Cx^* \leq c^u, \qquad\qquad (9.2a)$$

$$(Cx^* - c^l)_i y_i^* = 0, \quad i \in \mathscr{A}^l(x^*), \qquad y_i^* \geq 0, \quad i \in \mathscr{A}^l(x^*), \qquad (9.2b)$$

$$(Cx^* - c^u)_i y_i^* = 0, \quad i \in \mathscr{A}^u(x^*), \qquad y_i^* \leq 0, \quad i \in \mathscr{A}^u(x^*). \qquad (9.2c)$$

Conversely, every primal-dual pair (x^*, y^*) which satisfies conditions (9.2) is a global solution of QP (9.1) due to semidefiniteness of the Hessian matrix B. The primal-dual solution is unique if and only if the following two conditions are satisfied:

1. The active constraint rows $C_i, i \in \mathscr{A}^e \cup \mathscr{A}^l \cup \mathscr{A}^u$, are linearly independent.
2. Matrix B is positive definite on the null space of the active constraints.

9.1.2 Existing methods

All existing methods for solving QPs are iterative and can be grossly divided into Active Set and Interior Point methods. Interior Point methods are sometimes called Barrier methods due to the possibility of different interpretations of the resulting subproblems, see, e.g., Nocedal and Wright [121]. As a defining feature, Active Set methods keep a working set of active constraints and solve a sequence of equality constrained QPs. The working set must be updated between the iterates according to exchange rules which are based on conditions (9.2). In contrast, Interior Point methods do not use a working set but follow a nonlinear path, the so-called central path, from a strictly feasible point towards the solution.

Active Set methods can be divided into primal, dual, and parametric methods, of which the primal Active Set method is the oldest and can be seen as a direct extension of the Simplex Algorithm [36]. Dual Active Set methods apply the primal Active Set method to the dual of QP (9.1) (which exists if B is semidefinite). A relatively new variant of Active Set methods are Parametric Active Set Methods (PASM), e.g., the PQP method due to Best [17], which are the methods of interest in this thesis. PASMs are based on an affine-linear homotopy between a QP with known solution and the QP to be solved. It turns out that the optimal solutions depend piecewise affine-linear on the homotopy parameter and that along each affine-linear segment the active set is constant. The iterates of the method are simply the start points of each segment.

The numerical behavior of Active Set and Interior Point methods is usually quite different: While Active Set methods need on average substantially more iterations than Interior Point methods, the numerical effort for one iteration is substantially less for Active Set methods. Often one or the other method will perform favorably on a certain problem instance, underlining that both approaches are important.

We want to concisely compare the main advantages of the different Active Set versus Interior Point methods. One advantage of Interior Point methods is the regularizing effect of the central path which leads to well-defined behavior on problems with nonunique solutions due to, e.g., degeneracy or zero curvature in a feasible direction at the solution. An advantage of all Active Set methods is the possibility of hot starts which can give a substantial speed-up when solving a sequence of related QPs because the active set between the solutions usually changes only slightly. A unique advantage of PASM is that the so-called Phase 1 is not needed. The term Phase 1 describes the solution of an auxiliary problem to find a feasible starting point for primal and dual Active Set methods or a strictly feasible starting point for Interior Point methods. The generation of an appropriate starting point with Phase 1 can be as expensive as the subsequent solution of the actual problem.

9.1.3 Existing software

The popularity of primal/dual Active Set and Interior Point methods is reflected in the large availability of free and commercial software products. A detailed list and comparison of the various implementations is beyond the scope of this thesis. We restrict ourselves to citing a few implementations which we consider popular in Table 9.1. We further restrict our list to implementations which are specifically designed for QPs, although any NLP solver should be able to solve QPs.

The packages GALAHAD and FICO$^{(TM)}$ Xpress also provide the possibility of using crossover algorithms which start with an Interior Point method to eventually refine the solution by few steps of an Active Set method. CPLEX additionally

Code/Package	Interior Point	primal/dual Active Set	Parametric Active Set
BPMPD [114]	+		
BQPD [55]		+	
COPL_QP [168]	+		
CPLEX [89]	+	+	
CVXOPT [35]	+		
GALAHAD [69]	+	+	
HOPDM [62]	+		
HSL [6]	+	+	
IQP [21]		+	
LOQO [157]	+		
MOSEK [116]	+		
OOQP [58]	+		
qpOASES [52]			+
QPOPT [60]		+	
QuadProg++ [46]		+	
quadprog [112]		+	
QuadProg [153]		+	
rpasm [129]			+
Xpress Optim. Suite [54]	+	+	

Table 9.1: Software for convex Quadratic Programming (in alphabetical order).

offers the option of a concurrent optimizer which starts a Barrier and an Active Set method in parallel and returns the solution which was found in the shortest amount of CPU time.

For Parametric Active Set methods we are only aware of the code qpOASES (see Ferreau [52], Ferreau et al. [53]). We have developed a prototype Matlab$^{®}$ code called rpasm to demonstrate the efficacy of the proposed techniques and strategies to fortify reliability of PASM.

9.2 Parametric Active Set methods

9.2.1 The Parametric QP

The idea behind Parametric Active Set methods consists of following the optimal solutions on a homotopy path between two QP instances. Figuratively speaking, the homotopy morphs a QP with known solution into the QP to be solved. Let this homotopy be parametrized by $\tau \in [0,1]$. We then want to solve the one-parametric family of τ-dependent QP problems

$$\underset{x(\tau) \in \mathbb{R}^n}{\text{minimize}} \frac{1}{2} x(\tau)^{\mathrm{T}} B x(\tau) + b(\tau)^{\mathrm{T}} x(\tau) \quad \text{s.t.} \quad c^{\mathrm{l}}(\tau) \leq C x(\tau) \leq c^{\mathrm{u}}(\tau), \quad (9.3)$$

with b, c^{l}, and c^{u} now being continuous functions $b(\tau), c^{\mathrm{l}}(\tau), c^{\mathrm{u}}(\tau)$. For fixed τ, let the optimal primal-dual solution be denoted by $z(\tau) = (x(\tau), y(\tau))$ which necessarily satisfies (9.2) (with $b = b(\tau), c^{\mathrm{l}} = c^{\mathrm{l}}(\tau), c^{\mathrm{u}} = c^{\mathrm{u}}(\tau)$). If we furthermore restrict the homotopy to affine-linear functions $b \in \mathcal{H}^n, c^{\mathrm{l}}, c^{\mathrm{u}} \in \mathcal{H}^m$, where

$$\mathcal{H}^k = \{ f : [0,1] \to \mathbb{R} \mid f(\tau) = (1-\tau)f(0) + \tau f(1), \quad \tau \in [0,1] \},$$

it turns out that the optimal solutions $z(\tau)$ depend piecewise linearly but not necessarily continuously on τ (see Best [17]). On each linear segment the active set is constant. Parametric Active Set algorithms follow $z(\tau)$ by jumping from one beginning of a segment to the next. We can immediately observe that this approach allows hot-starts in a natural way. As mentioned already in Section 9.1.2, no Phase 1 is needed to begin the method: We can always recede to the homotopy start $b(0) = 0, c^{\mathrm{l}}(0) = 0, c^{\mathrm{u}}(0) = 0, x(0) = 0, y(0) = 0$, although this is certainly not the best choice as we discuss in Section 9.3 and Section 9.4.

9.2.2 The Parametric Quadratic Programming algorithm

A Parametric Active Set method was described by Best [17] under the name Parametric Quadratic Programming (PQP) algorithm. Algorithm 2 displays the main steps. The lines preceded by a number deserve further explanation.

Step 1: Computation of step direction. The working set W is encoded as an m-vector with entries 0 or ± 1, where the i-th component indicates whether constraint i is inactive ($W_i = 0$), active at the lower bound ($W_i = -1$), or active at the upper bound ($W_i = +1$). Let C_W denote the matrix consisting of the rows C_i with $W_i \neq 0$ and let $c_W(\tau)$ denote a vector which consists of entries $c_i^{\mathrm{l}}(\tau)$ or $c_i^{\mathrm{u}}(\tau)$ depending

---

Algorithm 2: The Parametric Quadratic Programming Algorithm.

Data: $B, C, b(1), c^l(1), c^u(1)$, and $z(0) = (x(0), y(0))$ optimal for $b(0), c^l(0), c^u(0)$
with working set $W \in \{-1, 0, 1\}^m$
Result: $z(1) = (x(1), y(1))$ or *infeasible* or *unbounded*
$\tau := 0$;

1 Compute step direction $\Delta z = (\Delta x, \Delta y)$ with current working set W;
2 Determine maximum homotopy step $\Delta\tau$;
   if $\Delta\tau \geq 1 - \tau$ **then** return solution $z(1) := z(\tau) + (1 - \tau)\Delta z$;
   Set $\tau^+ := \tau + \Delta\tau, z(\tau^+) := z(\tau) + \Delta\tau\Delta z$, and $W^+ := W$;
   if *constraint l is blocking constraint* **then**
      Set $W_l^+ := \pm 1$;
3   Linear independence test for new working set W^+;
      if *linear dependent* **then**
4   Try to find exchange index k;
         if *not possible* **then** return *infeasible*;
5   Adjust dual variables $y(\tau^+)$;
         Set $W_k^+ := 0$;
      end
   else *(sign change of k-th dual variable is blocking)*
      Set $W_k^+ := 0$;
6   Test for curvature of B on new working set W^+;
      if *nonpositive curvature* **then**
7   Try to find exchange index l;
         if *not possible* **then** return *unbounded*;
8   Adjust primal variables $x(\tau^+)$;
         Set $W_l^+ := \pm 1$;
      end
   end
   Set $\tau := \tau^+$ and $W := W^+$;
9 Possibly update matrix decompositions;
   Continue with Step 1;

---

on which (if any) bound is marked active in W_i. We can then determine the step
direction $(\Delta x, \Delta y)$ by solving

$$K_W(\tau) \begin{pmatrix} \Delta x \\ -\Delta y_W \end{pmatrix} := \begin{pmatrix} B & C_W^T \\ C_W & 0 \end{pmatrix} \begin{pmatrix} \Delta x \\ -\Delta y_W \end{pmatrix} = \begin{pmatrix} -(b(1) - b(\tau)) \\ c_W(1) - c_W(\tau) \end{pmatrix}. \quad (9.4)$$

The dual step Δy must be assembled from Δy_W by filling in zeros at the entries of
constraints i which are not in the working set (i.e., $W_i = 0$). For the initial working

set W we assume matrix C_W to have full rank and matrix B to be positive definite on the null space of C_W. Thus matrix $K_W(0)$ is invertible. As we shall see in Steps 3 and 6, the PQP algorithm ensures the full rank and positive definiteness properties and thus invertibility of $K_W(\tau)$ for all further steps through exchange rules for the working set W. We shall discuss a null space approach for the factorization of $K_W(\tau)$ in Step 9.

Step 2: Determination of step length. We can follow $z(\tau)$ in direction Δz along the current segment until either an inactive constraint becomes active (blocking constraint) or until the dual variable of a constraint in the working set becomes zero (blocking dual variable). Following the straight line with direction Δz beyond this point would lead to violation of conditions (9.2). The step length $\Delta\tau$ can be determined by *ratio tests*

$$\text{RT} : \mathbb{R}^{m+m} \to \mathbb{R} \cup \{\infty\}, \quad \text{RT}(u,v) = \min\{u_i/v_i \mid i \in \overline{m}, v_i > 0\}. \tag{9.5}$$

If the set of ratios is empty the minimum yields ∞ by convention. With the help of RT, the maximum step towards the first blocking constraint is given by

$$t_p = \min\{\text{RT}(Cx(\tau) - c^l, -C\Delta x), \text{RT}(c^u - Cx(\tau), C\Delta x)\}, \tag{9.6}$$

and towards the first blocking dual variable by

$$t_d = \text{RT}(W \circ y(\tau), W \circ \Delta y), \tag{9.7}$$

where $\circ$ denotes elementwise multiplication to compensate for the opposite signs of the dual variables for lower and upper active constraints. The maximum step allowed is therefore

$$\Delta\tau = \min\{t_p, t_d\}.$$

Best [17] assumes that each occurring minimization yields either ∞ or a unique minimizer with corresponding index $l \in \overline{m}$ if $\Delta\tau = t_p$ or index $k \in \overline{m}$ if $\Delta\tau = t_d$ from the sets of the ratio tests. The occurrence of at least one nonunique minimizer is called a *tie*. We can distinguish between primal-dual ties if $t_p = t_d$, primal ties if l is not unique, and dual ties if k is not unique. In case of a tie it is not clear which constraint should be added or removed from the working set W and bad choices can lead to cycling or even stalling of the method. Thus successful treatment of ties is paramount to the reliability of Parametric Active Set methods and shall be further discussed in Section 9.4.3.

Step 3: Linear independence test. The addition of a new constraint l to the working set W can lead to rank deficiency of C_{W+} and thus loss of invertibility of matrix $K_W(\tau^+)$. The linear dependence of C_l on $C_i, i : W_i \neq 0$ can be verified by solving

$$\begin{pmatrix} B & C_W^T \\ C_W & 0 \end{pmatrix} \begin{pmatrix} s \\ \xi_W \end{pmatrix} = \begin{pmatrix} C_l^T \\ 0 \end{pmatrix}. \tag{9.8}$$

Only if $s = 0$ then C_l is linearly dependent on $C_i, i : W_i \neq 0$ (see Best [17]). The linear independence test can be evaluated cheaply by reusing the factorization needed to solve the step equation (9.4).

Step 4: Determination of exchange index k. It holds that $s = 0$. Let ξ be constructed from ξ_W like Δy from Δy_W. Equation (9.8) then yields

$$C_l = \sum_{i:W_i \neq 0} \xi_i C_i. \tag{9.9}$$

Multiplying equation (9.9) by λW_l^+ with $\lambda \geq 0$ and adding this as a special form of zero to the stationarity condition in equations (9.2) yields

$$\begin{aligned} B(x(\tau) + \Delta\tau\Delta x) - b(\tau^+) &= \sum_{i:W_i \neq 0} y_i(\tau^+) C_i^T \\ &= -\lambda W_l^+ C_l^T + \sum_{i:W_i \neq 0} (y_i(\tau^+) + \lambda W_l^+ \xi_i) C_i^T. \end{aligned} \tag{9.10}$$

Thus all coefficients of $C_i, i : W_i^+ \neq 0$ on the right hand side of equation (9.10) are also valid choices $\tilde{y}$ for the dual variables as long as they satisfy the sign condition $W_i^+ \tilde{y}_i \leq 0$. Hence we can compute the largest such λ with the ratio test

$$\lambda = \mathrm{RT}(-W_l^+ (W \circ y(\tau^+)), W_l^+ (W \circ \xi)). \tag{9.11}$$

If $\lambda = \infty$ then the parametric QP does not possess a feasible point beyond τ^+ and thus the QP to be solved (at $\tau = 1$) is infeasible. Otherwise, let k be a minimizing index of the ratio set.

Step 5: Jump in dual variables. Now let

$$\tilde{y}_i = \begin{cases} -\lambda W_i^+ & \text{for } i = l, \\ y_i(\tau^+) + \lambda W_i^+ \xi_i & \text{for } i : W_i \neq 0, \end{cases}$$

and set $y(\tau^+) := \tilde{y}$. It follows from construction of λ that $\tilde{y}_k = 0$ and thus, constraint k can leave the working set. As a consequence, matrix $C_{W^+\setminus\{k\}}$ preserves the full rank property and has the same null space as C_W, thus securing regularity of matrix $K_{W^+}(\tau^+)$.

Step 6: Curvature test. The removal of a constraint from the working set can lead to exposure of directions of zero curvature on the null space of C_{W^+} (which is larger than the null space of C_W) leading to singularity of matrix $K_{W+}(\tau^+)$. Singularity can be detected by solving

$$\begin{pmatrix} B & C_W^T \\ C_W & 0 \end{pmatrix} \begin{pmatrix} s \\ \xi_W \end{pmatrix} = \begin{pmatrix} 0 \\ -(e_k)_W \end{pmatrix}, \qquad (9.12)$$

where e_k is the k-th column of the m-by-m identity matrix. Only if $\xi = 0$ then B is singular on the null space of C_{W+} (see Best [17]). As for the linear independence test of Step 3, the curvature test can be evaluated cheaply by reusing the factorization needed to solve the step equation (9.4).

Step 7: Determination of exchange index l. It holds that $\xi = 0$ and s solves

$$Bs = 0, \quad C_k s = -1, \quad C_{W+}s = 0. \qquad (9.13)$$

Therefore all points $\tilde{x} = x(\tau^+) + \sigma s$ are also solutions if $\tilde{x}$ is feasible. We can compute the largest such $\sigma = \min\{\sigma^l, \sigma^u\}$ with the ratio tests

$$\sigma^l = \mathrm{RT}(Cx(\tau^+) - c^l, -Cs), \quad \sigma^u = \mathrm{RT}(c^u - Cx(\tau^+), Cs). \qquad (9.14)$$

If $\sigma = \infty$ then the parametric QP is unbounded beyond τ^+, including the QP to be solved (at $\tau = 1$). Otherwise, let l be a minimizing index of a ratio set which delivers a final minimizer of σ.

Step 8: Jump in primal variables. Now set $x(\tau^+) := x(\tau^+) + \sigma s$. By construction of σ we have that either $C_l x(\tau^+) = c_l^l$ (if $\sigma = \sigma^l$) or $C_l x(\tau^+) = c_l^u$ (otherwise). Thus l can be added to the working set via $W_l^+ := -1$ (if $\sigma = \sigma^l$) or $W_l^+ := +1$.

Step 9: Update matrix decompositions. We summarize a null space approach for the solution of systems (9.4), (9.8), and (9.12). A range space approach is in general not possible if B is only semidefinite (see, e.g., Nocedal and Wright [121]). A direct factorization of $K_W(\tau)$ via LDL^T factorization is also possible but update formulae are in the general case not as efficient as for the null space approach (see Lauer [101]). The alternative of iterative linear algebra methods for the indefinite matrix $K_W(\tau)$ are beyond the scope of this thesis.

The null space approach is based on a QR decomposition of the transposed active constraint matrix

$$C_W^T = Q\tilde{R} = \begin{pmatrix} Y & Z \end{pmatrix} \begin{pmatrix} R \\ 0 \end{pmatrix} = YR, \quad Q^T Q = \mathbb{I}.$$

Thus the columns of Z constitute an orthonormal basis of the null space of C_W. The columns of Y are an orthonormal basis of the range space of C_W^T and the upper triangular matrix R is invertible due to the full rank assumption on C_W. By assumption, the *projected Hessian* Z^TBZ is positive definite and lends itself to a Cholesky decomposition

$$Z^TBZ = LL^T$$

with invertible triangular matrix L. Exploiting $C_W Z = 0$ and $C_W Y = R$, the unitary basis transformation

$$\begin{pmatrix} Y & Z & 0 \\ 0 & 0 & \mathbb{I} \end{pmatrix}^T \begin{pmatrix} B & C_W^T \\ C_W & 0 \end{pmatrix} \begin{pmatrix} Y & Z & 0 \\ 0 & 0 & \mathbb{I} \end{pmatrix} = \begin{pmatrix} Y^TBY & Y^TBZ & R \\ Z^TBY & LL^T & 0 \\ R^T & 0 & 0 \end{pmatrix}$$

yields a block-triangular system which can be solved via backsubstitution. When the working set W changes by addition, removal, or substitution of constraints, the QR decomposition of C_{W+} and following the Cholesky decomposition can be updated cheaply from the previous decompositions (see Gill et al. [59]). For exploitation of special structures of B and C in large scale applications we refer the interested reader to Kirches et al. [97, 96].

This concludes our presentation of the Parametric Quadratic Programming algorithm.

9.2.3 Far bounds

Many applications lead to QPs where some of the constraint bounds $c_i^l, c_j^u, i \neq j$, are infinite to allow for one-sided constraints, e.g., $0 \leq x_i \leq \infty$. However, a homotopy from finite to infinite $c^l(\tau)$ and $c^u(\tau)$ is not possible. The flipping bounds strategy to be described in Section 9.4.1 relies on finiteness of $c^l(\tau)$ and $c^u(\tau)$. We circumvent this problem by application of a so-called *far bounds* strategy. It is based on the following idea: Let $M > 0$ be given. If M is large enough then a solution (x,y) of (9.1) is also a solution of

$$\underset{x \in \mathbb{R}^n}{\text{minimize}} \frac{1}{2} x^T Bx + b^T x \quad \text{s.t.} \quad \tilde{c}^l \leq Cx \leq \tilde{c}^u, \tag{9.15}$$

where $\tilde{c}_i^l = \max(c_i^l, -M), \tilde{c}_i^u = \min(c_i^u, M), i = 1, \ldots, m$. We call a constraint bound which attains the value $\pm M$ a far bound. Algorithmically we solve a sequence of QPs with growing far bounds value M, see Algorithm 3. The total solution time will mostly be dominated by the solution of the first QP as consecutive QPs of the form (9.15) can be efficiently hot-started.

Algorithm 3: The far bounds strategy.

Initialize $M = 10^3$;
repeat
   | Solve QP (9.15);
   | **if** *no far bounds active* **then** return QP solution;
   | Grow far bounds: $M := 10^3 M$;
until $M > 10^{20}$;
if *last QP infeasible* **then** return QP infeasible;
else return QP unbounded;

9.3 Fundamental numerical challenges

In this section we describe the numerical challenges that occur in the PQP algorithm. We shall develop countermeasures in Section 9.4.

One fundamental challenge in many applications is ill-posedness of problems: Small changes in the data of the problem lead to large changes in the solution. This challenge necessarily propagates through the algorithm and leads to ill-conditioned matrices K_W. As a consequence the results of the step computation (9.4), the linear independence test (9.8), and the curvature test (9.12) can be erroneous up to $\text{cond}(K_W)$ times machine precision in relative error (see, e.g., Wilkinson [165]). This, in turn, can lead to very instable ratio tests (9.5) and wrong choices for the working set which can cause the algorithm to break down.

Rounding errors can also accumulate over several iterations and lead to the parametric "solution" $z(\tau)$ being optimal with an accuracy much less than machine precision. We call this phenomenon *drift*. Large drift can also lead to breakdown of the algorithm because the general assumption of optimality of $z(\tau)$ is violated.

Furthermore, the termination criterion must be adapted to work reliably on both well- and ill-conditioned problems.

Ill-conditioning can also be introduced if the null space of C_W captures two eigenvalues of B with high ratio, leading to ill-conditioning of the Cholesky factors L. In the extreme case, the next step for the dual variables can be afflicted with a large error, causing again instability in the ratio tests.

The second fundamental challenge is the occurrence of comparisons with zero, a delicate subject in the presence of rounding errors. These comparisons permeate the algorithm from the sign condition in the ratio tests (9.5) to the tests for linear dependence (9.8) or zero curvature (9.12).

The third fundamental challenge is the treatment of ties, i.e., nonuniqueness of minimizers of the ratio tests (9.5). Consider the case mentioned in Section 9.2.1

of a homotopy starting at $x(0) = 0, y(0) = 0, b(0) = 0, c^{\mathrm{l}}(0) = 0, c^{\mathrm{u}}(0) = 0, W = 0$. Clearly $(x(0), y(0))$ is an optimal solution at $\tau = 0$ regardless of the choice of B and C. Note that for the PQP algorithm the choice of $W = 0$ is only possible if B is positive definite. The first step direction will then point towards the unconstrained minimizer of the objective. If more than one constraint is active in the solution at $\tau = 1$ then the primal ratio test (9.6) for determination of the step length yields a (multiple) primal tie with $\Delta\tau = t^{\mathrm{p}} = 0$. Of all possible ratio test minimizers, one has to be chosen. One approach seems to be to employ pricing heuristics from primal/dual Active Set methods but we prefer a different approach which we discuss in Section 9.4.3. In the following iterations primal-dual ties can occur while still $\Delta\tau = 0$. Thus *cycling*, the repeated addition and removal of the same constraints without any progress, can be possible which leads to stalling of the method. We are not aware of any pricing strategy which can avoid the problem of cycling. Ties also occur naturally in the case of degenerate QPs, where the optimal primal or dual variables are not uniquely determined.

9.4 Strategies to meet numerical challenges

We propose to employ the following strategies for Parametric Active Set methods to meet the fundamental numerical challenges described in Section 9.3.

9.4.1 Rounding errors and ill-conditioning

The most effective countermeasure against the challenges of ill-conditioning is to iteratively improve the quality of the linear system solutions via

Iterative Refinement (see, e.g., Wilkinson [165]). We have already mentioned that the relative error in the solution z of

$$K_W z = d$$

can be as high as $\mathrm{cond}(K_W)$ times machine precision. Thus if $\mathrm{cond}(K_W) \approx 10^{10}$ and we perform computations in double precision, the solution z can have as little as six valid decimal digits. Iterative Refinement

$$z^0 = 0, \quad r^k = K_W z^k - d, \quad K_W \delta z^k = r^k, \quad z^{k+1} = z^k - \delta z^k$$

recovers a fixed number of extra valid digits in each iteration. In the previous example, the iterate z^2 has at least twelve valid decimal digits after only one extra step of Iterative Refinement. It is worth noticing that compared to the original "solution" z^1 each iteration only needs to perform one additional matrix-vector-multiplication with K_W and one backwards solve with the decomposition of K_W described in Section 9.2.2. In exact arithmetic $z^{k+1} = z^k$ for all $k \geq 1$.

Drift correction. A very effective strategy to avoid drift can be formulated if the PQP algorithm is cast in a slightly different framework. After each iteration, we rescale the homotopy parameter to $\tau = 0$, thus interpreting the iterate $z(\tau^+)$ as a new starting value $z(0)$. This does not avoid drift yet but allows for modifications to restore consistency of the starting point via

$$c_i^l(0) := \begin{cases} C_i x(0) & \text{if } W_i = -1, \\ \min\{c_i^l(0), C_i x(0)\} & \text{otherwise,} \end{cases}$$

$$c_i^u(0) := \begin{cases} C_i x(0) & \text{if } W_i = +1, \\ \max\{c_i^u(0), C_i x(0)\} & \text{otherwise,} \end{cases}$$

$$y_i(0) := \begin{cases} \max\{0, y_i(0)\} & \text{if } W_i = -1, \\ \min\{0, y_i(0)\} & \text{if } W_i = +1, \\ 0, & \text{otherwise,} \end{cases}$$

$$b(0) := Bx(0) - C^T y(0),$$

for $i \in \overline{m}$. This annihilates the effects of drift after every iteration, at the cost of splitting up the single homotopy into a sequence of homotopies which are, however, very close to the remaining part of the original homotopy. In exact arithmetic the proposed modification does not alter any value.

Termination Criterion. It is tempting to use the homotopy parameter τ in the termination criterion as proposed in Algorithm 2. However, this choice renders the termination criterion dependent on the choice of the homotopy start, an undesirable property. Instead we propose to use the relative distance δ in the data space

$$\Delta^0 = (b(\tau), c^l(\tau), c^u(\tau)), \qquad \Delta^1 = (b(1), c^l(1), c^u(1)),$$
$$s_j = (\Delta_j^1 - \Delta_j^0)/\max\{1, |\Delta_j^1|\}, \qquad j = 1, \ldots, n+m+m, \qquad \delta = \|s\|_\infty.$$

This choice also renders the termination criterion independent of the condition number of $K_W(1)$. We observe that the termination criterion can give no guarantee for the distance to the exact solution. Instead a backwards analysis result holds: The computed solution is the exact solution to a problem which is as close as δ to the one to be solved. The numerical results presented in Section 9.6 were obtained with the tolerance $\delta \leq \delta_{\text{term}} = 1e7\,\text{eps}$.

Ill-conditioning of the Cholesky factors L. To avoid ill-conditioning of the Cholesky factors L we have developed the so-called *flipping bounds* strategy. Flipping bounds is similar to taking long steps in the dual Simplex method (see Kostina [98], Sager [137]), where one variable changes in the working set from upper to lower bound immediately without becoming inactive in between, i.e., it *flips*. Flipping is only possible if $c^l(1)$ and $c^u(1)$ have only finite entries, which is guaranteed by the far bounds strategy described in Section 9.2.3. We modify the PQP algorithm in the following way: If a constraint l was removed without requiring another constraint k to enter the active set, we monitor the size of the smallest entry ℓ_i of the diagonal of L in Step 9. If $\ell_i^2 < \delta_{\mathrm{curv}}$ we have detected a small eigenvalue in L which corresponds to a small eigenvalue of B now uncovered through the removal of constraint l. To avoid ill-conditioning of LL^T, we introduce a jump in the QP homotopy by requiring that the other bound of constraint l is moved such that it becomes active immediately (hence the name flipping bounds) through setting

$$
\begin{aligned}
\tilde{c}_l^l(\tau^+) &:= c_l^u(\tau^+), W_l^+ = -1, && \text{if } W_l = +1,\\
\tilde{c}_l^u(\tau^+) &:= c_l^l(\tau^+), W_l^+ = +1, && \text{if } W_l = -1.
\end{aligned}
$$

The entries $j \neq l$ of $\tilde{c}^l$ and $\tilde{c}^u$ are set to the corresponding entries of c^l and c^u. Consequently the Cholesky decomposition from the previous step stays valid for the current projected Hessian.

The numerical results presented in Section 9.6 were obtained with the curvature tolerance $\delta_{\mathrm{curv}} = 1\mathrm{e}4\,\mathrm{eps}$.

9.4.2 Comparison with zero

Ratio tests. In the ideal ratio test (9.5) we take a minimum over a subset of m quotients with strictly positive denominator. The presence of round-off error makes it necessary to substitute the ideal ratio test by an expression with adjustable tolerances, e.g.,

$$
u_i^{\mathrm{cut}} = \max(u_i, \varepsilon_{\mathrm{cut}}), \quad i \in \overline{m},
$$
$$
\mathrm{RT}_r(u, v, \varepsilon_{\mathrm{cut}}, \varepsilon_{\mathrm{den}}, \varepsilon_{\mathrm{num}}) = \min\{u_i^{\mathrm{cut}}/v_i \mid i \in \overline{m}, v_i \geq \varepsilon_{\mathrm{den}}, u_i^{\mathrm{cut}} \geq \varepsilon_{\mathrm{num}}\}.
$$

We now explain the purpose of the three tolerances: The *denominator tolerance* $\varepsilon_{\mathrm{den}} > 0$ describes which small but positive values of v_i should already be considered less than or equal to zero. They are consequently discarded as candidates for the minimum.

The *cutting tolerance* $\varepsilon_{\mathrm{cut}}$ and the *numerator tolerance* $\varepsilon_{\mathrm{num}}$ offer the freedom of two different treatments for numerators close to zero. If $\varepsilon_{\mathrm{cut}} > \varepsilon_{\mathrm{num}}$ then negative numerators are simply cut off at ε_{cut} before the quotients are taken, yielding that the minimum is greater or equal to $\varepsilon_{cut}/\varepsilon_{den}$. For instance, we set $\varepsilon_{cut} = 0$ in the ratio tests for determination of the step length (9.6) and (9.7). This choice is motivated by the fact that in exact arithmetic $u_i \geq 0$ for all $i \in \overline{m}$ with $v_i > 0$. Thus only values u_i which are negative due to round-off are manipulated and the step length satisfies $\Delta\tau \geq 0$ also in finite precision arithmetic.

If $\varepsilon_{cut} \leq \varepsilon_{num}$ then cutting does not have any effect. We have found it beneficial for the reliability of PASM to set $\varepsilon_{num} = \varepsilon_{den}$ in the ratio tests (9.11) and (9.14) for finding exchange indices.

The numerical results presented in Section 9.6 were obtained with the ratio test tolerances $\varepsilon_{den} = -\varepsilon_{num} = 1e3\,eps$ and $\varepsilon_{cut} = 0$ for step length determination and $\varepsilon_{den} = \varepsilon_{num} = -\varepsilon_{cut} = 1e3\,eps$ for the remaining ratio tests.

Linear independence and zero curvature test. After solution of systems (9.8) and (9.12) for s and ξ_W we must compare the norm of s or ξ with zero. Let $\zeta = (s, \xi)$. We propose to use the relative conditions

$$\|s\|_\infty \leq \varepsilon_{test} \|\zeta\|_\infty \qquad \text{for the linear dependence test and} \qquad (9.16)$$

$$\|\xi\|_\infty \leq \varepsilon_{test} \|\zeta\|_\infty \qquad \text{for the zero curvature test.} \qquad (9.17)$$

We remark that $\|\zeta\|_\infty = \|\xi\|_\infty$ if $s = 0$ and $\|\zeta\|_\infty = \|s\|_\infty$ if $\xi = 0$. Thus we can replace $\|\zeta\|_\infty$ in the code by $\|\xi\|_\infty$ in test (9.16) and by $\|s\|_\infty$ in test (9.17). The numerical results presented in Section 9.6 were obtained with $\varepsilon_{test} = 1e5\,eps$.

9.4.3 Cycling and ties

Once ties have occurred, their resolution is a costly affair because of the combinatorial nature of the decision which subset of the possible constraints should be chosen to leave or enter the working set. This decision can be based on the solution of yet another QP of larger size than the original problem (see Wang [160]) or on heuristics similar to anti-cycling rules in the Simplex method.

We prefer a different approach instead. The idea behind the strategy we propose for ties is simple: Instead of trying to treat ties, we try to avoid them in the first place. The strategy is as simple as the idea and exploits the homotopy framework of PASM. Let a homotopy start $b(0), c^l(0), c^u(0)$ with optimal solution $(x(0), y(0))$ and working set W be given. Then for every triple of m-vectors $r^0, r^1, r^2 \geq 0$ the primal-dual pair $(x(0), \bar{y}(0))$ with

$$\bar{y}_i(0) = \begin{cases} y_i(0) + r_i & \text{if } W_i = -1, \\ y_i(0) & \text{if } W_i = 0, \\ y_i(0) - r_i & \text{if } W_i = +1, \end{cases} \qquad i \in \overline{m},$$

is an optimal solution to the homotopy start $\tilde{b}(0), \tilde{c}^l(0), \tilde{c}^u(0)$, where for $i \in \overline{m}$

$$\tilde{c}_i^l(0) = \begin{cases} c_i^l(0), & \text{if } W_i = -1, \\ c_i^l(0) - r_i^1, & \text{otherwise,} \end{cases}$$

$$\tilde{c}_i^u(0) = \begin{cases} c_i^u(0), & \text{if } W_i = +1, \\ c_i^u(0) + r_i^2, & \text{otherwise,} \end{cases}$$

$$\tilde{b}(0) = -(Bx(0) - C^T \tilde{y}(0)).$$

In other words, if we move the inactive constraint bounds further away from $Cx(0)$ and the dual variables of the active constraints further away from zero, $x(0)$ stays feasible and $b(0)$ can be adapted to restore optimality of $(x(0), \tilde{y}(0))$ with the same working set W. Recall that the ratio tests depend exactly on the residuals of the inactive constraints and the dual variables of the active constraints. In our numerical tests, the simple choice of

$$r_i^j = (1 + (i-1)/(m-1))/2, \quad j = 0, 1, 2, \quad i \in \overline{m},$$

has proved to work reliably. Because of the shape of r^j, we call this strategy *ramping*. It is important to avoid two entries of r^j to have the same value because many QP problems exhibit special structures, e.g., variable bounds of the same value for several variables which lead to primal ties if the homotopy starts with the same value for each of these variables. Of course, the choice of linear ramping is somewhat arbitrary and if a problem happens to have variable bounds in the form of a ramp, ties are again possible. However, this kind of structure is far less common than equal variable bounds.

We employ ramping in the starting point of the homotopy and also after an iteration which resulted in a zero step $\Delta \tau = 0$. Of course, this can lead to large jumps in the problem homotopy and practically catapult the current $b(0) := \tilde{b}(0)$ further away from $b(1)$. However, a PASM is capable of reducing even a large distance in the data space to zero in one step, provided the active set is correct. Thus the distance of the working set W to the active set of the solution is a more appropriate measure of the progress of a PASM. By construction, the active set is preserved by the ramping strategy.

We further want to remark that ties can never be completely avoided. For instance in case of a QP whose solution lies in a degenerate corner, a tie must occur in (at least) one iteration of a PASM. In the numerical examples we have treated so far, the ramping strategy effectively deferred these ties to the final step, where a tie is not a problem any more because the solution at the end of the last homotopy segment is already one of infinitely many solutions of the QP to be solved and no ties must be resolved in the solution.

9.5 The code rpasm: A PASM in Matlab®

We have implemented the strategies proposed in Section 9.4 in a Matlab® code called rpasm. The main purpose of the code is to demonstrate reliability and solution quality on the test set. In favor of code simplicity we have refrained from employing special structure exploiting linear algebra routines which could further enhance the runtime of the code. The three main features in the C++ PASM code qpOASES (see Ferreau [52], Ferreau et al. [53]) for runtime improvement in the linear algebra routines are special treatment of variable bounds, updates for QR decompositions, and appropriate updates for Cholesky decompositions. Of the three, rpasm only performs QR updates. Variable bounds are simply treated as general inequality constraints. Cholesky decompositions are computed from scratch after a change in the active set. Another feature which is common in most commercial QP solvers is the use of a preprocessing/presolve step to reduce the problem size by eliminating fixed variables and dispensable constraints and possibly scaling the data. We shall see that rpasm works reliably even without preprocessing.

9.6 Comparison with existing software

From the codes contained in Table 9.1 we use the ones which are freely available for academic purposes and come with a Matlab® interface, i.e., CPLEX, OOQP, qpOASES, plus the Matlab® solver quadprog and the newly developed rpasm. The programs cover the range of Primal Active Set (CPLEXP, quadprog), Dual Active Set (CPLEXD), Barrier/Interior Point (CPLEXB, OOQP), and Parametric Active Set (qpOASES, rpasm). For rpasm, we further differentiate between a version without iterative refinement (rpasm0) and with one possible step of iterative refinement (rpasm1). All codes were used with their default settings on all problems.

9.6.1 Criteria for comparison

We compare the runtime and the quality of the solution. Runtime was measured as the average runtime of three runs on one core of an Intel® Core™ i7 with 2.67 GHz and 8 MB cache in Matlab® 7.6 under Linux 2.6 (64 bit). The quality of solutions (x^*, y^*) was measured using a residual ρ of conditions (9.2) defined via

$$\rho_{\text{stat}} = \left\| Bx^* + b - C^T y^* \right\|_\infty,$$

$$\rho_{\text{feas}} = \max(0, c^l - Cx^*, Cx^* - c^u),$$

$$\rho_{\text{cmpl}}^l = \max\{ \left| (Cx^* - c^l)_i y_i^* \right| \mid y_i^* \geq +10\,\text{eps}\},$$

$$\rho_{\text{cmpl}}^u = \max\{ \left| (Cx^* - c^u)_i y_i^* \right| \mid y_i^* \leq -10\,\text{eps}\},$$

$$\rho = \max(\rho_{\text{stat}}, \rho_{\text{feas}}, \rho_{\text{cmpl}}^l, \rho_{\text{cmpl}}^u).$$

We visualize the results for problems from the Maros-Mészáros test set [111] with at most $n = 1000$ variables and $m = 1001$ two-sided inequality constraints (not counting variable bound constraints) in the performance graphs of Figures 9.1 and 9.2. The graphs display a time factor on the abscissa versus the percentage of problems that each code was able to solve within the time factor times the run-time of the fastest method for each problem. Roughly speaking, the graph of a fast method is close to the left hand side of the diagram, the graph of a reliable method is close to the top of the diagram. We remark that the results for rpasm were obtained using only dense linear algebra routines.

   There is a certain arbitrariness in the notion of a "solved problem" between the different codes. We choose to consider a problem as solved if ρ is less than or equal to a certain threshold. This approach is not unproblematic either: A not tight enough termination threshold of a code can lead to premature termination and the problem would be considered "not solved" by our criterion, although the method might have been able to recover a better solution with more iterations. This is especially an issue for Interior Point/Barrier methods. Thus the graphs in Figures 9.1 and 9.2 show reliability of the methods only in connection with their default settings. However, we are not aware of any simple procedure which would lead to a fairer comparison. Figure 9.1 shows the results with a relatively loose threshold of $\rho \leq$ 1e-2 and Figure 9.2 with a tighter threshold of $\rho \leq$ 1e-8.

9.6.2 Discussion of numerical results

We first discuss the results depicted in Figure 9.1 and continue with the differences to the tighter residual tolerance in Figure 9.2.

   From Figure 9.1 we see that the newly developed code rpasm with iterative refinement is the only code which solves all of the problems to the prescribed accuracy. The version of rpasm without iterative refinement fails on three problems (95 %). Furthermore, both versions of rpasm dominate quadprog both in

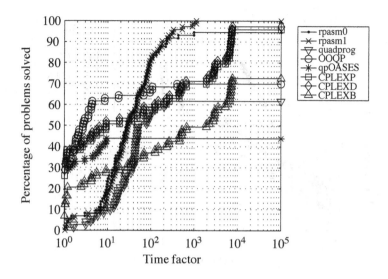

Figure 9.1: Performance comparison with loose residual threshold $\rho \leq$ 1e-2.

runtime and the number of solved problems (62 %). The primal and dual versions of CPLEX are the second most reliable with 96 % and 97 %. CPLEX solves no problem in less than 1.3 s, not even the small examples which are solved in a few milliseconds by rpasm. We suspect that this is due to a calling overhead in CPLEX, e.g., for license checking. This is also one reason why OOQP is much faster than the Barrier version of CPLEX, albeit they both solve roughly the same number of problems (70 % and 73 %, respectively). Even though the code qpOASES is only appropriate for QPs with positive definite Hessian, which make up only 27 % of the considered problems, it still solves 44 % of the test problems. Additionally, we want to stress that those problems solved by qpOASES were indeed solved quickly.

Now we discuss the differences between Figure 9.2 and Figure 9.1, i.e., when switching to a tighter residual tolerance of $\rho \leq$ 1e-8: The ratio of solved problems drops dramatically for the Interior Point/Barrier methods (CPLEXB: 29 %, OOQP: 37 %). This is a known fact and the reason for the existence of crossover methods which refine the results of Interior Point/Barrier methods with an Active Set method. The code qpOASES still solves 44 % of the problems, which indicates that the solutions that qpOASES yields are of high quality. Furthermore,

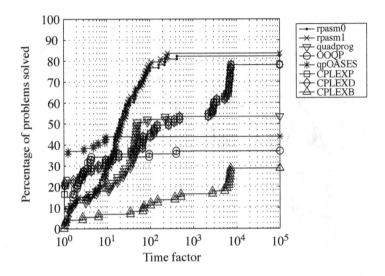

Figure 9.2: Performance comparison with tight residual threshold $\rho \leq$ 1e-8.

qpOASES is fast: It solves 36 % of the problems within 110 % of the time of the fastest method for each of these problems. The number of problems solved by quadprog decreases to 53 %. The primal and dual Active Set versions of CPLEX solve 78 % of the problems. Only the code rpasm is able to solve more than 80 % of the problems to a residual of $\rho \leq$ 1e-8 (rpasm0: 82 %, rpasm1: 84 %).

We can conclude that the strategies proposed in Section 9.4 indeed lead to a reliable method for the solution of convex QPs.

9.7 Drawbacks of the proposed PASM

Although the method has proved to work successfully on the test set, the improvement in reliability is achieved only at the price of breaking the pure homotopy paradigm which complicates an otherwise straightforward proof of convergence for the method: Drift correction, ramping, and the flipping bounds strategy lead to jumps in the trajectories of $b(\tau), c^l(\tau)$, and $c^u(\tau)$ and thus to a sequence of (possibly nonphysical) homotopies. Proving the nonexistence or possibility of cycles caused by these strategies is future work.

9.8 Nonconvex Quadratic Programs

The flipping bounds strategy presented in Section 9.4.1 can also be extended to the case of nonconvex QPs with indefinite Hessian matrix B. When the Cholesky factorization or update breaks down due to a negative diagonal entry, we also flip instead of remove the constraint l. Hence the projected Hessian always stays positive definite. By the second order necessary optimality condition, the projected Hessian in every isolated local minimum of the nonconvex QP is guaranteed to be positive semi-definite. Conversely, if the projected Hessian is positive definite and strict complementarity holds at $\tau = 1$ we obtain a local minimum because the second order sufficient condition is satisfied. No guarantees can be given in the case of violation of strict complementarity.

Finding a global minimum of a nonconvex QP is known to be an NP-hard problem, even if the Hessian has only one single negative eigenvalue (see Murty [118]). However, a local solution returned by the PASM can be refined by flipping all combinations of active bounds whose removal would lead to an indefinite projected Hessian and restarting the PASM for each of these flipped Active Sets, revealing again the combinatorial nature of finding the global solution of the nonconvex QP.

In the context of SQP with indefinite Hessian approximations (e.g., symmetric rank one updates, the exact Hessian, etc.), a local solution of a nonconvex QP is sufficient because the SQP method can only find local minima anyway.

For proof of concept we seek a local solution of the nonconvex problem

$$\text{minimize } \frac{1}{2}\sum_{i=1}^{k-2}\left(x_{k+i+1}-x_{k+i}\right)^2 - \frac{1}{2}\sum_{i=1}^{k=1}\left(x_{k-i}+x_{k+i}+\alpha_{k-i+1}\right)^2 \qquad (9.18a)$$

$$\text{s.t. } x_{k+i}-x_{i+1}+x_i = 0, \qquad\qquad\qquad i=1,\ldots,k-1, \qquad (9.18b)$$

$$\alpha_i \le x_i \le \alpha_{i+1}, \qquad\qquad\qquad\qquad i=1,\ldots,k, \qquad (9.18c)$$

$$0.4(\alpha_{i+2}-\alpha_i) \le x_{k+i} \le 0.6(\alpha_{i+2}-\alpha_i), \, i=1,\ldots,k-1, \qquad (9.18d)$$

with given constants $\alpha_i = 1 + 1.01^i, i = 1,\ldots,k-1$. We have adapted problem (9.18) from problem class 3 by Gould [64] by switching the sign in front of the second sum and the α terms in the objective. We start the computation with an initial guess of $x(0) = 0, y(0) = 0$, set the lower bounds of equations (9.18c) and (9.18d) active in the initial working set, and adjust the variables via ramping (see Section 9.4.3). The changes of the working set are depicted in Figure 9.3 for $k = 100$ and therefore $n = 199, m = 298$. Row l of the depicted image corresponds to the working set in iteration l and column j corresponds to the status of constraint j in the working set when the iterations advance. The shades indicate constraints which are inactive (gray), active at the lower bound (black), or active at the upper bound (white). Thus direct transitions from black to white or vice versa along a vertical line indicate flipping bounds. We can observe that the chosen initial working set is completely different to the final working set in the solution. Still the number of iterations is less than two times the number of constraints which indicates that the proposed method works efficiently on this instance of the nonconvex problem (9.18).

In the solution which corresponds to Figure 9.3, $n = 199$ out of m constraints are active and strict complementarity is satisfied. Thus we indeed have obtained a local optimum.

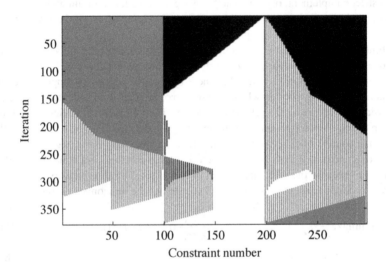

Figure 9.3: Active set changes for nonconvex problem (9.18), $k = 100$. Each line of the image corresponds to the working set in one iteration. The colors indicate constraints which are inactive (gray), active at the lower bound (black), or active at the upper bound (white). Direct transitions from black to white or vice versa along a vertical line indicate flipping bounds.

10 Automatic derivative generation

The inexact SQP method which we describe in Chapter 5 requires first and second order derivatives of the problem functions. There are several ways how derivatives can be provided. The first is to have them provided along with the problem-dependent model functions by the user. This can be cumbersome for the user and it is impossible for the program to check whether the derivatives are free of errors, even though consistency tests evaluated in a few points can somewhat mitigate the problem. These are, however, severe drawbacks.

A second way is the use of symbolic calculation of derivatives. Although in principle possible by the use of symbolic computer algebra systems, the resulting expressions for the derivatives can become too large to be evaluated efficiently.

A third way is the use of numerical schemes. Finite differences can be computed efficiently but they inevitably involve cancellation and truncation errors. While the cancellation errors can be circumvented by using a complex step derivative (see Squire and Trapp [148]) two further drawbacks still remain: First, complex step derivatives without cancellation are limited to first order derivatives. Second, the evaluation of the gradient of a scalar-valued function of many variables cannot be carried out efficiently.

The aim of this chapter is to recapitulate an efficient and automated way to compute derivatives from a given computer code. This fourth way does not suffer from the drawbacks of the previous three. The main principles are Algorithmic Differentiation (AD) and Internal Numerical Differentiation (IND). We refer the reader to Griewank [70] and Bock [22, 23], respectively.

The chapter is structured in four sections. In Section 10.1 we give a concise survey about the idea behind AD and in Section 10.2 about the principle of IND. We continue with the discussion of a subtle difficulty in the application of the IND principle to implicit time-stepping methods with monitor strategy in Section 10.3 and conclude the chapter in Section 10.4 with a short note on the numerical effort needed for the first and second order derivative generation needed in the inexact SQP method for NLP (3.3) described in Chapter 5.

10.1 Algorithmic Differentiation

Every computer code that approximates a mathematical function performs the calculation by concatenating a possibly large number of evaluations of a few *elemental operations* like $+, -, *, /, \sin, \exp$, etc., yielding an evaluation graph with intermediate results as vertices and elemental operations as edges. The principle of AD is to apply the chain rule to the concatenation of elemental operations. This is possible because the elemental operations are (at least locally) smooth functions.

There are two main ways how AD can be applied to compute derivatives of a function

$$F : \mathbb{R}^{n_\mathrm{ind}} \to \mathbb{R}^{n_\mathrm{dep}}$$

to machine precision.

Forward mode. We traverse the evaluation graph from the independent input variables towards the dependent output variables. The numerical effort of the forward mode to compute a directional derivative at $x \in \mathbb{R}^{n_\mathrm{ind}}$ in the direction of $s \in \mathbb{R}^{n_\mathrm{ind}}$

$$\nabla F(x)^{\mathsf{T}} s$$

is only a small multiple of the evaluation of $F(x)$.

Backward mode. We traverse the evaluation graph backwards from the dependent variables to the independent variables while accumulating the derivative information. The numerical effort of the backward mode to compute an adjoint directional derivative at $x \in \mathbb{R}^{n_\mathrm{ind}}$ in the direction of $s' \in \mathbb{R}^{n_\mathrm{dep}}$

$$\nabla F(x) s'$$

is also only a small multiple of the evaluation of $F(x)$. For the backward mode, however, the function $F(x)$ has to be evaluated in advance and all intermediate results must be accessible when traversing backwards through the evaluation graph. This is usually accomplished by storing all intermediate results to a contiguous memory block, the so called *tape*, or by a *checkpointing strategy* which stores only a few intermediate results and recomputes the remaining ones on the fly. Both approaches have their value depending on the ratio of computation speed and access time to the memory hierarchy on a particular computer architecture.

The elemental operations can be formally generalized to operate on truncated Taylor series. This approach makes the evaluation of arbitrary-order derivatives possible in a unified framework. To circumvent the drawback of high memory requirements and irregular memory access patterns, Griewank et al. [74] suggest to use only *univariate truncated Taylor series* from which mixed derivatives can be obtained by an interpolation procedure. Univariate Taylor coefficient propagation can also be used in forward and backward mode.

10.2 The principle of IND

For the solution of the NLP (3.3) we also need the derivatives of the state trajectories $\bar{u}^i$ and $\bar{v}^i$ with respect to initial values and controls on each multiple shooting interval which we denoted by the G-matrices and H-matrices in Chapter 8. For numerical efficiency reasons we compute the values of $\bar{u}^i$ and $\bar{v}^i$ with adaptive control of accuracy. Two problems arise for the computation of derivatives in this case:

First, if we consider the differential equation solver as a black box and employ finite differences or AD to the solver as a whole we inevitably also differentiate the adaptive components of the solver. This approach of External Numerical Differentiation (END) yields derivatives which are not consistent, i.e., in general they do not converge to the derivative of the exact solution when we increase the demanded accuracy. Even worse, the END-derivative is polluted with large errors if the adaptive components are not differentiable. This is rather the common case than the exception, e.g., if the adaptive components use conditional statements.

Second, we could try a differentiate-then-discretize approach: The derivative of the exact solution is given by a Variational Differential Equation (VDE) which exists in a forward and adjoint form (see, e.g., Hairer et al. [78], Hartman [79]). If only forward derivatives are needed then we can use a solver for the combined system of nominal and variational differential equations to obtain derivatives which are also consistent on a discretized level. However, if we apply a solver to the adjoint VDE we in general obtain a different discretization scheme due to adaptive error control. Thus the derivatives are not consistent on the discrete level which can severely impede the local convergence of the superordinate inexact SQP method.

IND solves these two problems. The principle of IND states:

1. The derivatives of an adaptive numerical procedure must be computed from the numerical scheme with all adaptive components kept constant (frozen).
2. The numerical scheme must be convergent for the nominal value and the derivative.

IND can be applied directly on the discrete level, e.g., by performing AD subject to skipping the differentiation of adaptive components, or indirectly, e.g., by choosing the same discretization scheme for the original and the variational differential equation.

10.3 IND for implicit time-stepping with monitor strategy

In this section we focus on a prototypical example which demonstrates the subtle issue of stability of the scheme for the VDE with implicit time-stepping methods. We restrict our presentation to the Backward Euler method although the results transfer to other implicit methods like Backward Differentiation Formula (BDF) methods with IND as described by Bauer et al. [13, 12], Bauer [11], Albersmeyer and Bock [3], Albersmeyer [2].

Example 5. Let us consider the linear heat equation on $\Omega = (0, \pi)$ with homogeneous Dirichlet boundary conditions

$$\partial_t u = \Delta u \quad \text{in } (0,1) \times \Omega,$$
$$u = 0 \quad \text{on } (0,1) \times \partial\Omega,$$
$$u\big|_{t=0} = u^0.$$

We discretize the problem with the FDM in space on the equidistant grid

$$x_j = jh, \quad j = 0, \ldots, N, \quad h = \pi/N,$$

and obtain the linear IVP

$$\dot{u}(t) = Au(t), \quad u(0) = u^0, \tag{10.1}$$

where the matrix A is given by

$$A = \frac{1}{h^2} \begin{pmatrix} -2 & 1 & & \\ 1 & \ddots & \ddots & \\ & \ddots & \ddots & 1 \\ & & 1 & -2 \end{pmatrix} \in \mathbb{R}^{(N-1)\times(N-1)}.$$

To satisfy the boundary conditions the values in the nodes x_0 and x_N are implicitly set to zero and are not a part of the discretized vector $u(t)$.

Lemma 10.1. *For $k = 1, \ldots, N-1$ define the pair $(v^k, \lambda_k) \in \mathbb{R}^{N-1} \times \mathbb{R}$ by*

$$v_j^k = \sin(jkh), \quad \lambda_k = 2h^{-2}(\cos(kh) - 1).$$

Then (v^k, λ_k) is an eigenpair of A.

Proof. Because $\sin(0kh) = 0$ and $\sin(Nkh) = \sin(k\pi) = 0$ we obtain

$$h^2(Av^k)_j = \sin(jkh - kh) + \sin(jkh + kh) - 2\sin(jkh)$$
$$= 2\sin(jkh)\cos(kh) - 2\sin(jkh)$$
$$= 2(\cos(kh) - 1)v_j^k,$$

for $j = 1,\ldots,N - 1$. This proves the assertion. $\square$

We see that the eigenvalue of smallest modulus is

$$\lambda_1 = 2h^{-2}(\cos(h) - 1) = 2h^{-2}\sum_{i=1}^{\infty}\frac{(-1)^i}{(2i)!}h^{2i} = -1 + \mathscr{O}(h^2),$$

and that the eigenvalue of largest modulus tends towards minus infinity

$$\lambda_{N-1} = \frac{2}{\pi^2}N^2(\cos(\pi(N-1)/N) - 1) \approx -\frac{4N^2}{\pi^2}.$$

Thus ODE (10.1) is stiff and becomes stiffer for finer spatial discretizations.

Because the vectors v^k are $N - 1$ eigenvectors to pairwise different eigenvalues λ^k they form a basis of $\mathbb{R}^{N-1}$. If we rewrite ODE (10.1) in the basis spanned by $\{v^k\}$ we obtain the decoupled ODE of Dahlquist type

$$\dot{\tilde{u}}(t) = \mathrm{diag}(\lambda_1,\ldots,\lambda_{N-1})\tilde{u}(t). \tag{10.2}$$

Consider now a Backward Euler method for ODE (10.2). Starting from $\tilde{u}^0$ at $t^0 = 0$ we compute a sequence $\{\tilde{u}^n\}$ such that the value $\tilde{u}^n$ at $t^n = t^{n-1} + \Delta t^n$ solves approximately

$$0 = (\mathbb{I} - \Delta t^n \,\mathrm{diag}(\lambda_1,\ldots,\lambda_{N-1}))\tilde{u}^n - \tilde{u}^{n-1} =: M^n\tilde{u}^n - \tilde{u}^{n-1}. \tag{10.3}$$

Although we could solve equation (10.3) exactly without much effort, this is not the case for nonlinear ODEs. Efficient numerical integrators employ a Newton-type method for the efficient solution of the nonlinear counterpart of equation (10.3). The so called *monitor strategy* is a Simplified Newton method for equation (10.3) where the decomposition of the iteration matrix M^k of a previous step is utilized and the contraction of the iteration is monitored. Usually we update the iteration matrix if the error has not been reduced satisfactorily within three Newton-type iterations. For simplification of presentation we assume that we perform enough iterations to reduce the error to machine precision. In the case of ODE (10.2) the iteration boils down to

$$\tilde{u}^{n,i} = \tilde{u}^{n,i-1} - (M^k)^{-1}(M^n\tilde{u}^{n,i-1} - \tilde{u}^{n-1}).$$

Thus we obtain for the j-th component of $\tilde{u}^{n,i}$ the expression

$$\tilde{u}_j^{n,i} = \left(1 - \frac{1 - \Delta t^n \lambda_j}{1 - \Delta t^k \lambda_j}\right)\tilde{u}_j^{n,i-1} + \frac{\tilde{u}_j^{n-1}}{1 - \Delta t^k \lambda_j}.$$

This iteration is unstable if and only if there exists an index j with $\tilde{u}_j^{n,0} \neq 0$ and such that

$$\left|1 - (1 - \Delta t^n \lambda_j)/(1 - \Delta t^k \lambda_j)\right| > 1,$$

or, equivalently,

$$\frac{1 - \Delta t^n \lambda_j}{1 - \Delta t^k \lambda_j} > 2 \quad \Leftrightarrow \quad 1 - \Delta t^n \lambda_j > 2 - 2\Delta t^k \lambda_j \quad \Leftrightarrow \quad \Delta t^n > \frac{1}{|\lambda_j|} + 2\Delta t^k. \quad (10.4)$$

A numerically optimal step size controller for stiff systems must maximize the step length such that the method is always on the verge of becoming unstable. We see from condition (10.4) that the step size can be more aggressively increased if the initial value does not contain modes which belong to eigenvalues of large modulus.

The following problem now becomes apparent: Assume that the initial value is a linear combination of only the first $p \ll N$ (low-frequency) modes, i.e.,

$$u^0 = \sum_{j=1}^{p} \tilde{u}_j^0 v^j.$$

Assume further that the monitor strategy keeps the iteration matrix at M^0 for the first steps and that the step size controller chooses

$$\Delta t^j = \frac{1}{|\lambda_{p+1}|} + 2\Delta t^1 < \frac{1}{|\lambda_p|} + 2\Delta t^1, \quad j = 2,\dots,n,$$

which yields a stable scheme. Say we want to compute the derivative of $u(t^n)$ with respect to u^0 in direction d by using the chosen scheme on the VDE for ODE (10.1), which is again ODE (10.1) with initial value d because ODE (10.1) is linear. If d has nonzero components in the directions of $v^k, k = p+2,\dots,N-1$, then this computation is not stable and thus not convergent. Hence this approach does not satisfy the IND principle.

One possible solution to this problem is to perform adaptive error control on the nominal values simultaneously with the forward derivative values (see, e.g., Albersmeyer [2, Section 6.7.5]). However, computing the gradient of the Lagrangian with forward sweeps is prohibitively expensive for large n_{ind}. Thus this approach is computationally not feasible for the problems considered in this thesis.

Our pragmatic approach, which has worked reliably for the examples in Part III, is to tighten the required tolerance for the Simplified Newton method that approximates the solution of the implicit system (equation (10.3) in our previous example) by a factor of 1000. This leads to more frequent reevaluation of the iteration matrix within the monitor strategy and enlarges thus the numerical stability regions. We have not yet performed a thorough comparison of the extra numerical effort of this approach which seems to amount to roughly 20%.

10.4 Numerical effort of IND

We have computed all numerical applications in Part III with the adaptive BDF method with IND derivative generation described by Albersmeyer [2]. The methods that he has developed enable us to evaluate matrix vector products of the form

$$J(x^k, y^k)v = \begin{pmatrix} \nabla_{xx}^2 \mathscr{L}(x^k, y^k) & -\nabla g(x^k) \\ \nabla g(x^k)^{\mathrm{T}} & 0 \end{pmatrix} \begin{pmatrix} v_1 \\ v_2 \end{pmatrix}$$

occurring in Chapter 5 in only a small multiple of the numerical effort spent for evaluation of $F(z^k)$. Even though the upper left block contains second derivatives of $\bar{u}^i$ and $\bar{v}^i$ they need only be evaluated in one adjoint direction given by the current Lagrange multipliers in y^k and the forward direction given by v_1.

11 The software package MUSCOP

We have implemented the inexact SQP method based on the GINKO algorithm with two-grid Newton-Picard generalized LISA as described in Chapter 5. It is our goal in this chapter to highlight the most important software design decisions and features. We have named the software package MUSCOP, which is an acronym for Multiple Shooting Code for PDEs. The name alludes to the successful software package MUSCOD-II (see Leineweber [105], Leineweber et al. [107] with extensions by Leineweber [106], Diehl [47], Schäfer [141], Sager [138]) because we had originally intended to design it as a MUSCOD-II extension. In Section 11.1 we discuss why we have decided to develop the method in a stand-alone parallel form and outline which programming paradigms have proved to be useful in the development of MUSCOP. Afterwards we explain the orchestration of the different software components in Section 11.2.

11.1 Programming paradigms

The programming paradigms of a software package should be chosen to support the main goals and target groups of the project. We identify two equally important target groups for MUSCOP: Practitioners and algorithm developers. Both groups have different perspectives on the goals of MUSCOP. In our opinion the main goals are:

1. Hassle-free setup of new application problems
2. Quick, accurate, and reliable solution of optimization problems
3. Fast development of algorithmical extensions

While goal (2) is of equal importance to both practitioners and developers, goal (1) will be more important than goal (3) for a practitioner and vice versa for a developer. A user of MUSCOP is in most real-life cases partly practitioner and developer at the same time.

11.1.1 Hybrid language programming

Quick problem solution and fast development of extensions sometimes are diametrically opposed goals: On the one hand the fastest runtimes might be achieved by only writing assembler machine code for a specific computer architecture, but such a code might soon become too complex and surpass a developer's capacity to maintain or extend it in a reasonable amount of time. On the other hand, the sole use of a high-level numerical programming language like Matlab$^{®}$ or GNU Octave might result in a considerable loss of performance, especially if algorithms are not easily vectorizable, while the development time of the code and its extensions might be dramatically reduced, mainly because debugging problems on a numerical level is possible in a more accessible way by the use of well-tested built-in methods like cond, eig, svd, etc., and data visualization.

We use hybrid language programming in the following way: All time-critical algorithmic components should be implemented in lower-level programming languages and all other components in higher-level programming languages. This concept is not new, GNU Octave being one example because it is written in C++ while most dense linear algebra components are based on an optimized BLAS and LAPACK implementation called ATLAS (see Whaley et al. [163]).

In the case of MUSCOP the most time-critical component is the evaluation of the shooting trajectories and their derivatives of first and second order (see Part III). This task is performed by the software package SolvIND (see Albersmeyer and Kirches [5]) which is entirely written in C++ and uses ATLAS, UMFPACK (see Davis [39]), and ADOL-C (see Griewank et al. [72, 73], Walther et al. [159]). Most of the remaining code is written in Matlab$^{®}$/GNU Octave except for the interface to SolvIND which is at the time of writing only available for GNU Octave and not for Matlab$^{®}$. This approach has proven to be beneficial for the development speed of MUSCOP while only minor performance penalties have to be accepted (see Part III).

The GNU Octave and C++ components of MUSCOP are separated on the left hand and right hand side of Figure 11.1, respectively. We shall give a more detailed explanation of Figure 11.1 in Section 11.2.

11.1.2 No data encapsulation

Figure 11.1 already indicates that the different software components of MUSCOP are heavily interwoven. This is not a result of poor design of programming blocks (or classes if you will). It is rather the inevitable consequence of intelligent structure exploitation in the MUSCOP algorithm. The efficiency of MUSCOP lies in the reuse of intermediate data of one logic program block in another one, which is a major efficiency principle not only but particularly in complex numerical methods.

Take for instance the use of the Newton-Picard Hessian approximation in Chapter 8, Section 8.4. It can only be assembled after half of the condensing steps, namely the computation of $\hat{Z}$, has been performed. Only then can we evaluate the partially projected (coarse grid) Hessian $\widetilde{B}' = \hat{Z}^T \hat{B} \hat{Z}$.

We do not want to suggest that a functional encapsulation in logical blocks like function evaluation, condensing, QP solution, etc. is impedimental. We believe, however, that the encapsulation of the data of these blocks is. In our opinion the structure of the code should indeed follow the logical structure of a mathematical exposition of the method but the exchange of data should not be artificially obstructed by interfaces which follow the functional blocks.

Object Oriented Programming (OOP) raises the rigid coupling of function and data interfaces (methods and private members in the language of OOP) to a design principle. We believe that OOP is a valid approach for software which is supposed to be used in a black-box fashion but we believe it to be more obstructive than helpful for the structure exploiting numerical methods we develop in this thesis. This is the main reason why MUSCOP is not OOP and not an extension of MUSCOD-II.

In MUSCOP we use a global, hierarchical data structure which is accessible in all software components, at least on the Matlab®/GNU Octave level. Hierarchical here means that the data structure consists of substructures which map the functional blocks to data blocks without hiding them. The biggest disadvantage of global variables is if course that users and developers have to know which variables they are allowed to write access and in what states the variables are when performing a read access. This task is without doubt a difficult one to accomplish. But the difficulty really stems from the complexity of the numerical method and not from the choice of computer language or programming paradigm. No programming paradigm can turn a complex and difficult-to-understand method into a simple and easy-to-understand code.

11.1.3 Algorithm centered not model centered

Another distinguishing design decision is that MUSCOP is centered around the GINKO Algorithm 1 of Chapter 5, in contrast to MUSCOD-II which is centered around the OCP model. This enables us to use the GINKO algorithm in MUSCOP also as a stand-alone inexact SQP method or LISA-Newton method. We shall describe in Section 11.2 how MUSCOP orchestrates the different software components around GINKO.

11.1.4 Reverse Communication interface

Reverse Communication seemed to be an antiquated method of interface design until it regained acceptance in the mid 90's within the linear algebra community

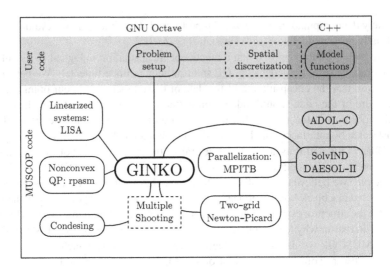

Figure 11.1: Schematic of the MUSCOP software architecture.

(see, e.g., Dongarra et al. [49], Lehoucq et al. [104]). Its most compelling features are simplicity and flexibility, especially when multiple programming languages are in use.

A program with Reverse Communication interface is called with an incomplete set of input data first. If more input data is needed the program returns to the caller indicating which input data is needed next. After computation of this data the user calls the program again passing the new input data. This procedure is iterated until the program signals termination to the user.

A typical example is an iterative linear algebra solver which returns to ask the user for a matrix vector multiplication or a preconditioner vector multiplication. Obviously the solver does not need to know the matrix or the preconditioner, nor does it need to pose restrictions on how they are represented.

GINKO also uses Reverse Communication. When called without input parameters GINKO initializes a data structure which is then modified by the user and passed back to GINKO. In this data structure there are two state flags, the *flow control* and the *action* flag. The flow control flag tells GINKO which part of the code is the next to be evaluated and the action flag tells the user on return which variables in the data structure must be freshly computed before GINKO can be called again. As a side note we want to remark that Reverse Communication is strongly coupled with a Finite State Machine programming paradigm.

The main advantage of Reverse Communication lies in the fact that GINKO does not pose any requirements on the form of function representations which allows for great flexibility and easy extensibility of the method. The disadvantage is that the developer has the responsibility to provide input data in a manner consistent with the method. But this is not a problem of programming but rather a problem of the numerical computing: The developer must know what he or she is doing (at least to a large extent).

11.2 Orchestration of software components

We now turn to the explanation of Figure 11.1 which depicts a schematic overview of the software components of MUSCOP and how they interact. As mentioned earlier the figure is divided into four areas: The lower area is MUSCOP code written in GNU Octave on the left hand side (white background) and in C++ on the right hand side (light gray background). The upper area depicts the user code written in GNU Octave on the left (light gray background) and C++ code on the right (dark gray background).

The two dashed boxes symbolize conceptual entities which do not necessarily have one block of code but are rather a placeholder to signify special structure in the data that flows along the connecting edges of the diagram. The *Spatial discretization* box is located over the border of the GNU Octave/C++ regions to indicate that the code for spatial discretization can be in either language (or even both).

11.2.1 Function and derivative evaluation

The model functions, in particular the discretized ODE and PDE right hand side functions $f^{ODE(l)}$ and $f^{PDE(l)}$, need to be evaluated many times and thus they are programmed in C++. We evaluate them via SolvIND either directly or via the IND integrator DAESOL-II (see Albersmeyer and Bock [3], Albersmeyer [2]).

SolvIND uses ADOL-C (see Walther et al. [159]) to automatically obtain first and second order derivatives of the model functions in forward and backward mode.

In MUSCOP we also take special care to extend the principle of IND to the function evaluations within the GINKO algorithm: When we evaluate the simplified Newton step in the monotonicity test we must freeze the integration scheme to obtain monotonicity on the discretized level. This feature avoids numerical pollution of the monotonicity test by effects of the adaptive error control of the integrator. When a step is accepted, we recompute the function values with a new adaptive scheme.

We parallelize the calls to the integrator on the Multiple Shooting structure so that the numerically most expensive part, the solution and differentiation of the local IVPs, can be evaluated concurrently. For parallelization we use the toolbox MPITB (see Fernández et al. [51]) which provides access to the message passing interface MPI (see, e.g., Gropp et al. [75]) from Matlab®/GNU Octave. Our manager-worker approach allows for parallel processing both on cluster computers and multi-core workstations. An advantage of computation on cluster computers is that the integration AD tapes can be stored locally and do not need to be exchanged between the cluster nodes, yielding an efficient memory parallelization without communication overhead.

The functions which need to be provided in GNU Octave code are loading of model libraries via SolvIND, grid prolongation and restriction, evaluation of variable norms suitable for the PDE discretization, visualization, and preparation of an initial solution guess.

11.2.2 Condensing and condensed QP solution

We carry out the solution of the large-scale QP subproblems via condensing and a PASM for the resulting small-scale to medium-scale QP (see Chapter 8). These QPs need to be solved within the generalized LISA (see Chapter 5). For the solution of the first QP in a major GINKO iteration, i.e., k loop, we need to compute the coarse grid Hessian and constraint Jacobian matrices. There is no need to reevaluate them for the following QP solutions until we increment k. We do need to reevaluate the fine grid Lagrange gradient and constraint residuals, plus one fine grid Hessian-vector product for the condensing procedure for each QP subproblem.

The QP solver rpasm (see Chapter 9) allows for efficient hot-starting of the first QP of a major GINKO iteration by reusing the working set of the previous iteration. MUSCOP then freezes the working set of the first QP for all following QPs until we increment k to commence the next major iteration.

If the working set of the previous iteration leads to a projected Hessian matrix which is not positive definite at the current iterate then we need to resort to a safe cold start of the PASM with trivial 0-by-0 projected Hessian.

11.2.3 Estimation of κ and ω

MUSCOP currently estimates κ in two ways: If the coarse grid and the fine grid are identical then GINKO is explicitly run with the information that $\kappa = 0$. This allows for performing only one step of LISA instead of the minimum of two steps required for an a-posteriori estimate of κ. If $\kappa = 0$ then one step of LISA already delivers the exact solution of the linearized system. If the fine grid does not coincide with the coarse grid we employ the Ritz κ-estimator.

The nonlinearity constant ω is implicitly estimated in the inexact Simplified Newton step via

$$\omega \left\| \delta z^k \right\| \approx [h_k^\delta]_* = \frac{2(1 - \bar{\rho}_{i+1}) \left\| \delta z_{i+1}^{k+1} - \delta z_0^{k+1} \right\|}{\alpha_k^2 \left\| \delta z^k \right\|}. \tag{11.1}$$

We observe that the right hand side of equation (11.1) is afflicted with a cancellation error in the norm of the numerator. This error is then amplified by α_k^{-2} causing that if the step size α_k drops below, say, 10^{-4} then $[h_k^\delta]_*$ might be overestimated. This in turn leads to even smaller step sizes

$$\alpha_k = \frac{1}{(1 + \rho)[h_k^\delta]_*}.$$

Thus GINKO gradually reduces α_k to zero and the method stalls. We have implemented an estimator for the cancellation error. The cancellation error is displayed for each iteration of MUSCOP but does not influence the iterations of MUSCOP. It rather serves as an indicator for failure analysis.

11.2.4 Automatic mesh refinement

In general the refinement is performed within the following framework: The user provides a conforming hierarchy of grids. MUSCOP starts with both coarse and fine grids on level $l = 0$. If the inexact Simplified Newton increment $\left\| \widetilde{\delta z^k} \right\|$ is smaller than a given threshold then we either terminate if level l is already the finest level or we increment l and use the prolongation of the current iterate as a starting guess for NLP (4.1) on the next level.

The variable steps on the following grid level can be used as a rough a-posteriori error estimation for the discretization error.

If in the course of computation GINKO signals that M^k needs to be improved because κ is too large then we automatically refine the coarse grid until κ is small enough again.

As mentioned earlier it will surely be advantageous to perform adaptive a-posteriori mesh refinement independently for the fine and the coarse grid and seperately for each shooting interval. This aspect, however, is beyond the scope of this thesis.

Part III

Applications and numerical results

12 Linear boundary control for the periodic 2D heat equation

We presents numerical results for the model problem (6.1) in this chapter. The computations have been published in Potschka et al. [131] and are given here for completeness.

This chapter is structured as follows: In Section 12.1 we list the problem parameters for the computations. Afterwards we discuss the effects of the Euclidean and the L^2 projector in the projective approximation of the Jacobian blocks in Section 12.2. In Section 12.3 we present numerical evidence for the mesh independence result of Theorem 6.7. We conclude this chapter with a comparison of the symmetric indefinite Newton-Picard preconditioners with a symmetric positive definite Schur complement preconditioner in a Krylov method setting in Section 12.4.

12.1 General parameters

The calculations were performed on $\Omega = [-1,1]^2$. We varied the diffusion coefficient $D \in \{0.1, 0.01, 0.001\}$ which results in problems with almost only fast modes for $D = 0.1$ and problems with more slow modes in the case of $D = 0.001$. The functions α and β were chosen identically to be a multiple of the characteristic function of the subset

$$\Gamma = \Gamma_1 \cup \Gamma_2 := (\{1\} \times [-1,1]) \cup ([-1,1] \times \{1\}) \subset \partial\Omega,$$

with $\alpha = \beta = 100\chi_\Gamma$. Throughout, we used the two boundary control functions

$$\bar{\psi}_1(x) = \chi_{\Gamma_1}(x), \quad \bar{\psi}_2(x) = \chi_{\Gamma_2}(x).$$

In other words, the two controls act each uniformly on one edge Γ_i of the domain.

With $\gamma = 0.001$, we chose the regularization parameter rather small such that the objective function is dominated by the tracking term which penalizes deviation of the state at the end of the period from the desired state $\hat{u}$. We used the discontinuous target function

$$\hat{u}(x) = \left(1 + \chi_{[0,1]\times[-1,0]}(x)\right)/2.$$

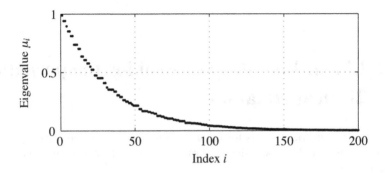

Figure 12.1: The eigenvalues μ_i of the spectrum of the monodromy matrix G_u decay exponentially fast. Only few eigenvalues are greater than 0.5. Shown are the first 200 eigenvalues calculated with $D = 0.01$ and $\beta = 100\chi(\Gamma)$ on a grid of 8-by-8 elements of order 5.

The controls were discretized in time on an equidistant grid of $m = 100$ intervals.

For the discretization of the initial state $u(0)$ we used quadrilateral high-order nodal Finite Elements. The reference element nodes are the Cartesian product of the Gauss-Lobatto nodes on the 1D reference element. We used part of the code which comes with the book of Hesthaven and Warburton [84], and extended the code with continuous elements in addition to discontinuous elements.

The evaluations of matrix-vector products with G_u and G_q were obtained from the Numerical Differentiation Formula (NDF) time-stepping scheme implemented in ode15s [146], which is part of the commercial software package Matlab®, with a relative integration tolerance of 10^{-11}. Due to the discontinuities in the controls, the integration was performed intervalwise on the control discretization grid. A typical spectrum of the monodromy matrix G_u can be seen in Figure 12.1. The approximations $\tilde{G}_u$ are calculated directly from the fundamental system projected on the slow modes or on the coarse grid.

Figure 12.2 shows the solution states and controls (u_0, q).

12.2 Euclidean vs. L^2 projector

Figure 12.4 summarizes the spectral properties of the iteration matrices $\tilde{J}^{-1}\Delta J$. The spectrum of the iteration matrix can also be interpreted as the deviation of the preconditioned system matrix from the identity. The discretization with 4-by-4

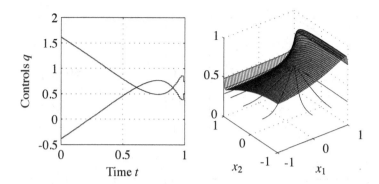

Figure 12.2: Optimal controls q (left) and optimal states u_0 (right) for target function $\hat{u}$, calculated for $D = 0.01$ on a grid of 32-by-32 elements of order 5. The displayed mesh is not the finite element mesh but an evaluation of the Finite Element function on a coarser equidistant mesh.

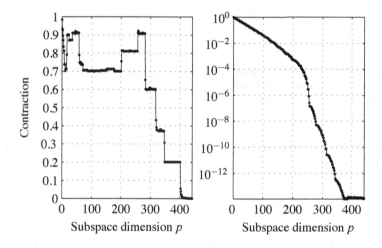

Figure 12.3: LISA contraction with Newton-Picard preconditioning versus the subspace dimension p for the Euclidean projector (left) and the L^2 projector. Note that the plot on the right hand side is in logarithmic scale.

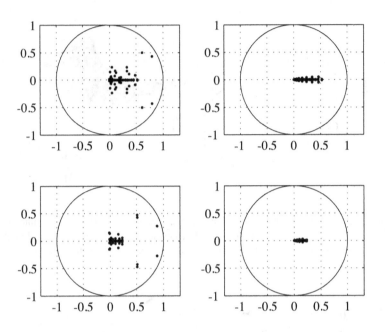

Figure 12.4: Top row: Unit circle and spectrum of iteration matrix for the classical Newton-Picard with $p = 20$ using Euclidean projector (left column) and L^2 projector (right column). Bottom row: Like top row with $p = 45$.

elements of order 5 is moderately fine in order to achieve reasonable computation times for the spectra.

Figures 12.3 and 12.4 depict that the appropriate choice of the projector for the Newton-Picard approximation leads to fast convergence which is monotonically decreasing in the dimension p of the slow subspace. In Figure 12.3 we see that both the Euclidean and the L^2 projector eliminate many large eigenvalues, but the Euclidean projector leaves out a few large eigenvalues which belong to eigenvectors which exhibit mesh-specific characteristics. Numerically we observe that the Euclidean projector leads to a non-monotone behavior of the contraction rate with respect to the subspace dimension, and also exhibits clear plateaus. The L^2 projector leads to an exponential decay of the contraction rate with respect to the subspace dimension and is by far superior to the Euclidean projector. Thus, only the L^2 projector will be considered further.

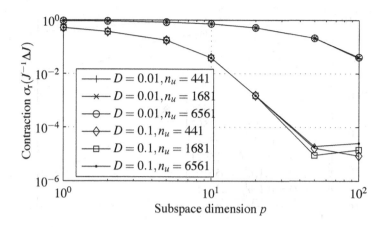

Figure 12.5: Asymptotic contraction rate for classical Newton-Picard preconditioning versus subspace dimension p for varying diffusion coefficient D and spatial degrees of freedom n_u.

12.3 Mesh independence

Figure 12.5 shows the asymptotic contraction rate of the iteration matrix $\tilde{J}^{-1}\Delta J$ of the basic linear splitting approach (6.2) with the classical Newton-Picard preconditioner for diffusion coefficients $D \in \{0.1, 0.01\}$ and spatial degrees of freedom $n_u \in \{441, 1681, 6561\}$ with respect to the subspace dimension. Figure 12.6 shows the same quantities for the two-grid version of the preconditioner for diffusion coefficients of $D = \{0.1, 0.01, 0.001\}$. We can observe that the contraction rate is independent of n_u in accordance with Corollary 6.8.

If we compare Figures 12.5 and 12.6 we see that the contraction for classical Newton-Picard is better than for two-grid Newton-Picard with subspace dimension $p = n_u^c$. However, better contraction is outweighed by the effort for constructing the dominant subspace spanned by V through IRAM already for rather small values of p. In particular for the case of $D = 0.001$, computation of V with $p > 10$ is prohibitively slow.

However, using a Krylov method like GMRES (see Saad and Schultz [136]) to accelerate the basic linear splitting approach (6.2) yields acceptable iteration numbers also for low values of p even though there is almost no contraction due to $\sigma_r(\tilde{J}^{-1}\Delta J) > 0.99$. For the extreme pure Picard case $p = 0$, we obtain a solution

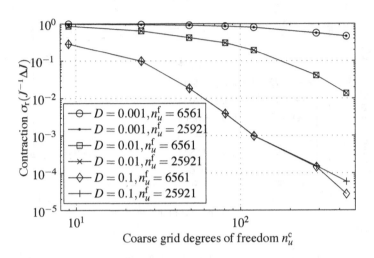

Figure 12.6: Asymptotic contraction rate for two-grid Newton-Picard preconditioning versus coarse grid degrees of freedom n_u^c for varying diffusion coefficient D and fine grid degrees of freedom n_u^f.

within $11, 34, 98$ iterations for $D = 0.1, 0.01, 0.001$, respectively, with a termination tolerance $\varepsilon_O = 10^{-4}$. We remark that for inexact inner solutions with $\varepsilon_M, \varepsilon_H$ much larger than machine precision, Flexible GMRES (see Saad [134]) should be employed.

As we have seen in Section 6.3.6, the effort on the coarse grid for the two-grid Newton-Picard preconditioner is negligible compared to the effort on the fine grid. Thus, even medium scale coarse grid degrees of freedom n_u^c are possible in practical computations and lead to fast contraction rates. In this case, acceleration of LISA (6.2) by nonlinear Krylov subspace methods does not lead to considerable savings in the number of iterations.

12.4 Comparison with Schur complement preconditioning

In Murphy et al. [117] it was shown that the symmetric positive definite exact Schur complement preconditioner

$$J_{\mathrm{MGW}} = \begin{pmatrix} M & 0 & 0 \\ 0 & N & 0 \\ 0 & 0 & (G_u - \mathbb{I}_{n_u})M^{-1}(G_u^{\mathrm{T}} - \mathbb{I}_{n_u}) + \gamma^{-1}G_q N^{-1}G_q^{\mathrm{T}} \end{pmatrix}$$

leads to $J_{\mathrm{MGW}}^{-1}\mathcal{J}$ having exactly three different eigenvalues 1 and $(1 \pm \sqrt{5})/2$. As a consequence, any Krylov subspace method with an optimality or Galerkin property converges within 3 iterations for the preconditioned system. Inversion of the lower right block of J_{MGW} is computationally prohibitively expensive but we can approximate this block by the Newton-Picard approach presented in Section 6.3 which leads with $\tilde{X} = (\tilde{G}_u - \mathbb{I}_{n_u})M^{-1}(\tilde{G}_u^{\mathrm{T}} - \mathbb{I}_{n_u}) \in \mathbb{R}^{n_u \times n_u}$ to the preconditioner

$$\tilde{J}_{\mathrm{MGW}} = \begin{pmatrix} M & 0 & 0 \\ 0 & N & 0 \\ 0 & 0 & \tilde{X} + \gamma^{-1}G_q N^{-1}G_q^{\mathrm{T}} \end{pmatrix}.$$

Now we can invoke again the Sherman-Morrison-Woodbury formula to obtain

$$\left(\tilde{X} + \gamma^{-1}G_q N^{-1}G_q^{\mathrm{T}}\right)^{-1} = \tilde{X}^{-1} - \tilde{X}^{-1}G_q(\gamma N + G_q^{\mathrm{T}}\tilde{X}^{-1}G_q)^{-1}G_q^{\mathrm{T}}\tilde{X}^{-1}$$
$$= \tilde{X}^{-1} - \tilde{X}^{-1}G_q H G_q^{\mathrm{T}}\tilde{X}^{-1},$$

with $\tilde{X}^{-1} = (\tilde{G}_u - \mathbb{I}_{n_u})^{-1}M(\tilde{G}_u^{\mathrm{T}} - \mathbb{I}_{n_u})^{-1}$. We observe that the occurring matrices coincide with the matrices which need to be inverted for the indefinite Newton-Picard preconditioner $\tilde{J}$ we have developed in Section 6.3. Thus, one iteration of an iterative method with $\tilde{J}_{\mathrm{MGW}}$ can be considered computationally as expensive as one iteration with $\tilde{J}$.

Because the preconditioner $\tilde{J}_{\mathrm{MGW}}$ is positive definite we can employ it within a symmetry exploiting Krylov subspace method like MINRES (see Paige and Saunders [123]), which is not possible with the indefinite preconditioner $\tilde{J}$. On the downside, it is not possible to use $\tilde{J}_{\mathrm{MGW}}$ in the basic linear splitting approach (6.2) because the real eigenvalues of the iteration matrix $\mathbb{I}_{n_1+n_2} - \tilde{J}_{\mathrm{MGW}}^{-1}\mathcal{J}$ cluster around 0 and $(1 \pm \sqrt{5})/2$. Since $(1 + \sqrt{5})/2 > 1$ LISA does not converge.

In Figure 12.7 we compare the number of iterations for symmetry exploiting MINRES preconditioned by $\tilde{J}_{\mathrm{MGW}}$ with the number of iterations for GMRES preconditioned by $\tilde{J}$ for varying fine and coarse grid degrees of freedom n_u^{f} and n_u^{c}.

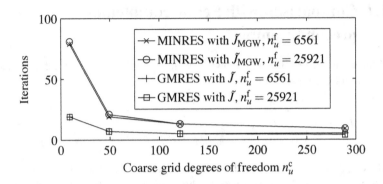

Figure 12.7: Comparison of the iterations of MINRES with Newton-Picard Schur complement preconditioner $\tilde{J}_{\mathrm{MGW}}$ and GMRES with the symmetric indefinite Newton-Picard preconditioner $\tilde{J}$ for varying fine and coarse grid degrees of freedom n_u^{f} and n_u^{c}.

We observe that the indefinite preconditioner $\tilde{J}$ is superior to $\tilde{J}_{\mathrm{MGW}}$ by a factor of
2–4 even though $\tilde{J}$ is not employed in a symmetry exploiting Krylov method.

    We remark that the indefinite preconditioning approach taken by Schöberl and
Zulehner [143] does not work in a straight forward way without an approximation
of the M-block in the preconditioner by a matrix $\hat{M}$ such that $\hat{M} - M$ is positive
definite. Thus, we do not include a comparison here.

13 Nonlinear boundary control of the periodic 1D heat equation

In this chapter we consider the problem of optimal nonlinear boundary control of the periodic heat equation

$$\underset{q\in L^2(\Sigma),u\in W(0,1)}{\text{minimize}} \quad \frac{1}{2}\int_\Omega (u(1;.)-\hat{u})^2 + \frac{\gamma}{2}\iint_\Sigma (q-\hat{q})^2 \tag{13.1a}$$

$$\text{s.t.} \quad \partial_t u = D\Delta u, \qquad \text{in } (0,1)\times\Omega, \tag{13.1b}$$

$$\partial_\nu u + \alpha u^4 = \beta q^4, \quad \text{in } (0,1)\times\partial\Omega, \tag{13.1c}$$

$$u(0;.) = u(1;.), \qquad \text{in } \Omega, \tag{13.1d}$$

on $\Omega = (0,1)$. We see that problem (13.1) is very similar to the model problem (6.1) except for the polynomial terms in the boundary control condition (13.1c) of Stefan–Boltzmann type.

13.1 Problem and algorithmical parameters

For our computations the desired state and control profiles are

$$\hat{u}(x) = 1 + \cos(\pi(x-1))/10, \quad \hat{q}(t,x) = 1.$$

The other problem parameters are given by

$$\gamma = 10^{-4}, \quad D = 1, \quad \alpha(t,0) = \beta(t,0) = 1, \quad \alpha(t,1) = \beta(t,1) = 0,$$

effectively resulting in a homogeneous Neumann boundary condition without control at $x = 1$. The control acts only via the boundary at $x = 0$.

We performed all computations with a relative integrator tolerance of 10^{-5}. The algorithm terminates if the primal-dual SQP step is below 10^{-4} in the suitable norm

$$\|z\| := \left(\left\|x^{\text{PDE}}\right\|_{\mathbb{I}\otimes M_V}^2 + \left\|x^{\text{rem}}\right\|_2^2 + \left\|y^{\text{PDE}}\right\|_{\mathbb{I}\otimes M_V^{-1}}^2 + \left\|y^{\text{rem}}\right\|_2^2 \right)^{1/2},$$

where for any symmetric positive definite matrix A we define $\|x\|_A^2 = x^T A x$ and where x^{PDE} denotes the composite vector of PDE states and y^{PDE} denotes the composite vector of dual variables for the PDE state dependent part of the time boundary constraint (3.3b) and the PDE continuity condition (3.3c). The variables x^{rem} and y^{rem} are placeholders for the remaining primal and dual variables. The occurrences of the mass matrices M_V in the Kronecker products (see Chapter 8) make sure that the PDE variables are measured in an L^2 sense. For the correct weighting of the dual PDE variables we have to consider that a dual PDE variable $\tilde{y}$ of NLP (4.1) is from the canonical dual space of $\mathbb{R}^{N_V}$. To obtain a discretized Riesz representation $\hat{y} \in \mathbb{R}^{N_V}$ in an L^2 sense we need to require that

$$\tilde{y}^T x = \hat{y}^T M_V x \quad \text{for all } x \in \mathbb{R}^{N_V}$$

and thus we obtain

$$\|\hat{y}\|_{M_V} = \left\| M_V^{-1} \tilde{y} \right\|_{M_V} = \|\tilde{y}\|_{M_V^{-1}}.$$

We performed the computations on a hierarchy of spatial FDM meshes with

$$N_V^l = 4 \cdot 8^{l-1} + 1$$

equidistant grid points on levels $l = 1, \ldots, 5$ and controls which are piecewise constant on grids of $n_{\mathrm{MS}} = 12, 24, 48$ equally sized intervals.

Figure 13.1 depicts the solution on the finest grid for the case of $n_{\mathrm{MS}} = 24$.

13.2 Discussion of numerical convergence

13.2.1 Grid refinements

We display the numerical self-convergence in Figure 13.2 for the case $n_{\mathrm{MS}} = 24$. The first 10 iterations are performed fully on grid level $l = 1$. In iterations 9–12 and starting again from iteration 16 on only full steps are taken ($\alpha_k = 1$). MUSCOP refines the fine grid for the first time after iteration 11 because the norm of the inexact Simplified Newton increment $\left\| \widetilde{\delta z}^{11} \right\|$ is small enough. Iteration 12 is the first iteration with $\hat{\kappa} > 0$. An estimate of $[\hat{\kappa}] = 1.21$ signals MUSCOP in iteration 13 that the coarse grid must be refined, too. The next refinements of the fine grid happen after iterations 18, 21, and 23. Only three iterations are performed on the finest grid level.

For the computations of the distances $\left\| z^k - z^* \right\|$ of the iterates z^k to the final solution approximation z^* we prolongate the iterates on coarser levels $l < 5$ to the finest level $l = 5$ and evaluate the norm on the finest level. We can observe

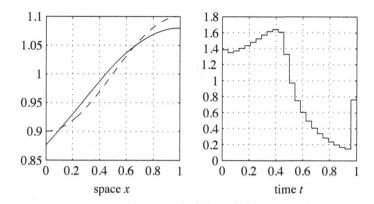

Figure 13.1: Solution of problem (13.1) with $n_{MS} = 24$. In the left panel we depict the state $u(1;.)$ at the period end (solid line) and the desired state $\hat{u}$ (dashed line). In the right panel we depict the optimal control q over one period.

that in contrast to the step norm $\|\delta z^k\|$ the error $\|z^k - z^*\|$ forms clear plateaus in iterations 10–12, 18–19, 21–22, and 23–24. These plateaus occur because the error in these iterations is dominated by the interpolation error of the spatial grid on the coarser levels. We thus suggest for efficiency reasons to couple the fine grid refinements to the contraction κ: If we can reduce the error by a factor of κ in one step then we should refine the grid such that the interpolation error is reduced with a similar magnitude. Thus we perform an aggressive refinement leading to eight times more grid points after each refinement. We observe a reduction in the (interpolation) error of about $1/8$ in Figure 13.2 between iterations 19–22 and 22–24. We can thus infer by extrapolation that the final error is dominated by the spatial interpolation error and lies around $7 \cdot 10^{-4}$. The observed error reduction of $\mathscr{O}(h)$ in the grid size h is optimal from a theoretical point of view because it coincides with the error of the spatial discretization.

We also perform refinement of the coarse grid aggressively. The rationale behind an efficient choice of the coarse grid is to choose the coarse grid fine enough to get fast convergence and thus fewer iterations on the numerically more expensive finer grids while maintaining moderate or negligible cost of the full derivatives on the coarse grid compared to the few directional derivatives on the fine grid.

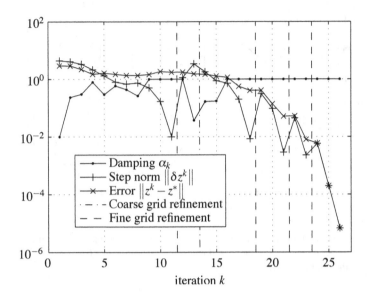

Figure 13.2: Self-convergence plot for problem (13.1).

13.2.2 Computation times

The computations were performed on up to four cores of an Intel® Core™ i7 with 2.67 GHz, 8 MB cache, and 18 GB RAM. In Table 13.1 we list the computation times of the different algorithmic parts of the code. We see that with 97.6 % most of the runtime is spent in the simulation and IND derivative generation of the dynamic systems. The evaluation and derivative generation for the non-dynamic functions, i.e., $\Phi^l, r^{b(1)}, r^i, r^e$, the solution of the QP subproblems, which comprises the condensing step for matrices and vectors, the PASM solution of the medium-scale QP, and the blow-up of the condensed solution to the uncondensed space all take up a negligible amount of time.

The numbers underline the hybrid programming approach that we have chosen. Most of the runtime is spent within the C++ code to generate solutions and derivatives of the dynamic systems.

The runtime can be reduced from 3289.3 s to 1714.9 s by exploitation of four cores. The resulting speedup of 1.9 is clearly suboptimal. There are two main reasons: First, each pair of the four cores shares one L2 cache and thus there are

Task	Time [s]	% of total
Simulation/IND	3209.8	97.6
Non-dynamic functions/AD	22.5	0.7
QP matrix condensing	6.2	0.2
QP vector condensing	3.7	0.1
QP solution	0.5	0.0
QP solution blow-up	5.9	0.2
Norm computations	1.7	0.1
Grid prolongations	6.7	0.2
Grid restrictions	15.5	0.5
GINKO	1.4	0.0
Remaining computations	15.4	0.5
Total	3289.3	100.0

Table 13.1: Timings for serial computation with $n_{MS} = 24$.

penalties in cache efficiency when running on four cores. Second, the adaptive timestepping results in different integration times on each shooting interval especially when on some intervals fast transients have to be resolved and on others not as we can observe in Figure 13.3. Such transients can for instance be caused by large jumps in the control variables (compare Figure 13.1). We have only implemented equal distribution of the shooting intervals to the different processes. This leads to many processes being idle until the last one has finished all its work. Optimal distribution of the processes is known as the *job shop scheduling problem.* One simple solution heuristic is, e.g., greedy work balancing which adaptively distributes the IVPs over the available processes by assigning the currently largest job to the first free process. The relative sizes of the jobs can be assumed to be known from the previous SQP step. In Figure 13.4 we can see that a greedy distribution leads to improved parallelism. A rigorous investigation of efficient parallelization is beyond the scope of this thesis.

13.2.3 Exact Hessian vs. two-grid approximation

In Tables 13.2 and 13.3 we compare the cumulative time spent in the simulation and IND of the dynamic systems for two different types of Hessian approximation:

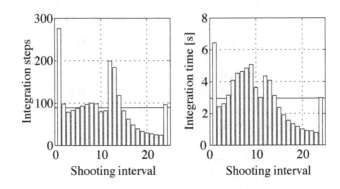

Figure 13.3: Steps and integration times per shooting interval on the finest level in the solution. The solid black lines indicate the average.

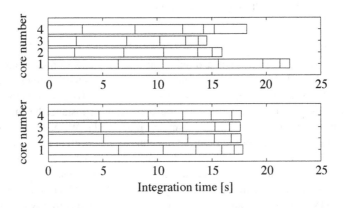

Figure 13.4: Comparison of regular distribution of IVPs to four processes/cores (upper) and greedy scheduling (lower). The termination time (makespan) can be significantly reduced.

The exact Hessian and the two-grid version (see Chapter 8). The quality of the two-grid Hessian approximation is so good that we obtain the solution after 25 major iterations in both cases. Usage of a two-grid approximation yields more evaluations of matrix vector products with the Hessian on the coarser grids but less on the finer grids. We observe that the two-grid Hessian approximation yields

Level l	Spatial N_V^l	Forward simulation	Jacobian MVP	Jacobian transpose MVP	Hessian MVP
1	5	11.5	2.3	2.9	40.8
2	33	24.7	15.0	7.6	110.7
3	257	11.9	3.2	9.0	151.6
4	2049	46.6	12.9	39.5	715.1
5	16385	584.8	187.1	498.6	10556.8

Table 13.2: Cumulative time [s] for simulation and IND on different mesh levels for exact Hessian approximation with $n_{MS} = 24$.

Level l	Spatial N_V^l	Forward simulation	Jacobian MVP	Jacobian transpose MVP	Hessian MVP
1	5	16.8	3.0	3.9	81.4
2	33	32.5	17.7	9.5	158.4
3	257	14.0	3.2	8.9	33.1
4	2049	54.3	12.9	39.5	151.5
5	16385	659.1	187.6	499.9	1994.7

Table 13.3: Cumulative time [s] for simulation and IND on different mesh levels for two-grid Hessian approximation with $n_{MS} = 24$.

a performance increase of 84 % for the Hessian evaluation on the finest grid. The overall wall-time savings on four cores amount to 68 % in this example.

13.2.4 Refinement of control in time

In Table 13.4 we present the number of SQP iterations on each spatial discretization level when we refine the control discretization in time. We observe that more iterations are needed for finer control discretizations but they are all spent on the coarsest levels $l = 1, 2$. The SQP iterations on the finer levels $l = 3, 4, 5$ coincide. The overall runtime for finer control discretizations increases due to three main reasons: First, the number of necessary grid transfer operations, i.e., prolongations and restrictions, increases linearly with number of shooting intervals n_{MS}.

Control time dis-	SQP iters on level $l =$					Runtime [s]		
cretization n_{MS}	1	2	3	4	5	QP	grid transfers	total
12	9	6	3	2	3	5.5	6.5	1042.4
24	11	7	3	2	3	20.7	19.0	1396.4
48	17	19	3	2	3	93.7	48.2	1842.0

Table 13.4: SQP iterations on each spatial discretization level and runtimes of selected parts for varying time discretizations of the control computed on four cores.

Level l	Spatial N_V^l	Forward simulation	Jacobian MVP	Jacobian transpose MVP	Hessian MVP
1	5	9.2	1.6	1.9	30.3
2	33	21.6	11.6	5.4	72.9
3	257	10.9	2.7	7.6	28.0
4	2049	41.6	10.8	33.3	126.7
5	16385	508.0	162.0	419.4	1714.9

Table 13.5: Cumulative time [s] for simulation and IND on different mesh levels for two-grid Hessian approximation with $n_{MS} = 12$.

Second, the effort for the solution of the QP subproblems increases because the amount of linear algebra operations for the condensing step (see Chapter 8) increases quadratically with n_{MS} and because the condensed QP grows linearly in size with n_{MS}. Third, we can see in Tables 13.5 and 13.6 that the effort for simulation and IND increases. The reason lies in the adaptivity of the IVP solver because every jump in the controls leads to transients in the dynamic system which require finer time steps to be resolved to the requested accuracy.

Level l	Spatial N_V^l	Forward simulation	Jacobian MVP	Jacobian transpose MVP	Hessian MVP
1	5	16.9	4.6	6.7	176.5
2	33	60.6	45.8	22.3	720.2
3	257	18.2	3.8	10.7	39.5
4	2049	68.6	15.2	46.6	178.0
5	16385	836.7	231.5	604.2	2417.2

Table 13.6: Cumulative time [s] for simulation and IND on different mesh levels for two-grid Hessian approximation with $n_{MS} = 48$.

14 Optimal control for a bacterial chemotaxis system

Chemotaxis is the phenomenon of single cells moving in a directed fashion in reaction to a chemical substance in their environment, e.g., to seek for food. In the seminal paper of Keller and Segel [95] a mathematical model for bacterial chemotaxis was proposed for the first time. For further information and bibliographical references see Horstmann [87, 88].

Chemotaxis can be explained by two phases of bacterial movement: A phase of tumbling movement similar to a random walk and a phase of directed movement through propulsion by flagella rotation (see Figure 14.1). The duration of each phase is controlled by a chemical substance in the environment, the so called *chemoattractant*. In environments with low chemoattractant concentration tumbling movement prevails while directed movement prevails in environments with higher chemoattractant concentration. The effect of this simple mechanism is that for large numbers of bacteria the bacteria will on average move upwards gradients of the chemoattractant concentration. This behavior leads to interesting dynamic phenomena like pattern formation and traveling waves of bacteria.

Figure 14.1: Simplified schematic of *E.coli* with rotating flagella for directed movement.

14.1 Problem formulation

We use the model of Tyson et al. [154, 155] which has been also used in a optimizing boundary control scenario by Lebiedz and Brandt-Pollmann [102] with the software MUSCOD-II. In contrast to their results, our approach allows for a much higher accuracy in the spatial discretization.

More specifically we consider the tracking type boundary control problem

$$\underset{z,c,q}{\text{minimize}} \quad \frac{1}{2}\int_\Omega (z(1,\cdot)-\hat{z})^2 + \frac{\gamma_c}{2}\int_\Omega (c(1,\cdot)-\hat{c})^2 + \frac{\gamma_q}{2}\int_0^1 q^2 \tag{14.1a}$$

$$\text{s.t.} \quad \partial_t z = D_z \Delta z + \alpha \nabla \cdot \left(\frac{\nabla c}{(1+c)^2} z \right), \qquad \text{in } (0,1)\times\Omega, \tag{14.1b}$$

$$\partial_t c = \Delta c + w\frac{z^2}{(\mu+z^2)} - \rho c \qquad \text{in } (0,1)\times\Omega, \tag{14.1c}$$

$$\partial_\nu z = 0, \qquad \text{in } (0,1)\times\partial\Omega, \tag{14.1d}$$

$$\partial_\nu c = \beta(q-c), \qquad \text{in } (0,1)\times\partial\Omega, \tag{14.1e}$$

$$z(0,.) = z_0, \tag{14.1f}$$

$$c(0,.) = c_0, \tag{14.1g}$$

$$q^u \geq q \geq q^l, \qquad \text{in } (0,1)\times\partial\Omega, \tag{14.1h}$$

where ∂_ν denotes the derivative in direction of the outwards pointing normal on Ω.

The objective (14.1a) penalizes the deviation of the cell density and the chemo-attractant concentration from given desired distributions at the end of the time horizon, and penalizes excessive use of the control. The governing system of PDEs is nonlinear. The difficulty lies in the chemotaxis term (preceded by α) of equation (14.1b) which is a convection term with nonlinear convection velocity $(1+c)^{-2}\nabla c$ for z. In equation (14.1c) we see that the chemoattractant evolves due to diffusion, is produced proportional to a nonlinear function of the cell density, and decays with a factor ρ. There is no flux of bacteria over the domain boundaries due to the Neumann condition (14.1d) but we can control the system via chemoattractant influx over the boundary in a Robin-type fashion according to condition (14.1e), where q describes a controllable chemoattractant concentration outside of the domain Ω. The optimization scenario (14.1) is not periodic in time. Instead, we prescribe initial values for the cell density and the chemoattractant concentration in equations (14.1f)–(14.1g). Finally we require the control q to be bounded between q^u and q^l.

14.2 Numerical results

We computed approximate solutions to the optimal control problem (14.1) with the problem data listed in Table 14.1. We used a four level hierarchy of spatial FDM grids with 17, 65, 257, and 1025 equidistant points for z and c and $n_{MS} = 36$ multiple shooting intervals. The computation ran 31 iterations in 15 min 40 s on

Symbol	Value	Description
Ω	$[0,1]$	spatial domain
$\hat{z}(x)$	$2x$	target cell density distribution
γ_c	0.5	weight for concentration tracking
$\hat{c}(x)$	0	target chemoattractant distribution
γ_q	1e-3	weight for control penalization
D_z	0.33	cell diffusion
α	80	chemotaxis coefficient
w	1	chemoattractant production coefficient
μ	1	chemoattractant production denominator
ρ	0	chemoattractant decay coefficient
β	0.1	Robin boundary control coefficient
$z_0(x)$	1	initial cell density distribution
$c_0(x)$	0	initial chemoattractant distribution
q^{u}	0.2	upper control bound
q^{l}	0.0	lower control bound

Table 14.1: Chemotaxis boundary control model data.

four cores. The IVP integrator performed between 19 and 64 integration steps per shooting interval with an average of 27.1 steps in the solution on the finest grid. Figure 14.2 shows a self-convergence plot. We observe that after refinement of the fine and the coarse grid, the globalization strategy needs to recede to damped steps for iterations 15 and 16. Afterwards only full steps are taken. Only four iterations are performed on the finest grid level with derivatives generated on the second coarsest level. The error plateaus are even more prominent than for the example in Chapter 13. From extrapolation of the error at iterations 15, 23, and 27, we can expect the final solution to be accurate to about $\left\| z^k - z^* \right\| \approx 0.1$. This accuracy is not satisfactory but on our current system and with our implementation, finer spatial discretizations are not possible. We do not believe that this is a generic problem of our approach because the main memory problem is that DAESOL-II currently keeps all IND tapes (including right hand side Jacobians and their decompositions) in main memory. This takes up the largest amount of memory within GINKO. We are positive that this memory bottleneck can be circumvented by previsional swapping out and in of tape entries to hard disk or by checkpointing techniques.

From Figure 14.2 we also see that many iterations on coarser grids, namely the iterations on the plateaus, can be saved if a reliable estimator for the interpolation error is available. For efficiency reasons the fine grid should be refined as soon as the inexact Simplified Newton increment $\left\| \widetilde{\delta z^k} \right\|$ is below the interpolation error of the spatial grid. This aspect is, however, beyond the scope of this thesis.

Moreover, if we consider the error reduction between iterations 21–22, 25–26, 29–30, and 30–31 we observe that the error reduction in the last iterates on each of the three finest grid levels is roughly (fine) grid independent.

We depict the optimal states at different snapshots in time in Figure 14.4 and the corresponding controls in Figure 14.3. We observe that in order to achieve the unnatural linear distribution $\hat{z}$ the optimal solution consists of a control action on the left boundary at the beginning, followed by a control action on the right boundary shortly afterwards. This effects the formation of two cell heaps close to the boundary. Finally a control action on the right boundary makes the left heap of cells move to the middle of the domain and the right heap to grow further towards the target cell distribution $\hat{z}$.

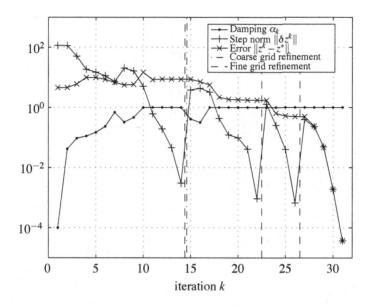

Figure 14.2: Self convergence plot for the chemotaxis problem (14.1).

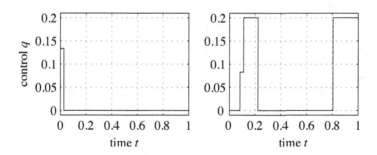

Figure 14.3: Optimal control profiles for the chemotaxis problem (14.1). The left hand panel shows the control at the boundary $x = 0$ and the right hand panel at $x = 1$.

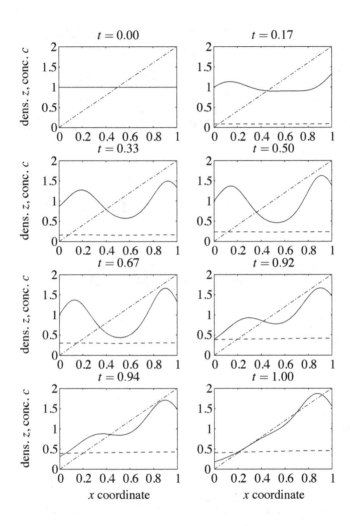

Figure 14.4: Optimal states for the chemotaxis problem (14.1). For different time points t we plot the bacteria density z (solid line) the chemoattractant concentration c (dashed line) and the bacteria target distribution (dash-dotted line).

15 Optimal control of a Simulated Moving Bed process

In this chapter we describe a variant of the Simulated Moving Bed (SMB) process. For completeness we quote in large parts from the article Potschka et al. [130].

In a chromatographic column, different components that are dissolved in a liquid are separated due to different affinities to the adsorbent. As a result the different components move with different velocities through the column and hence can be separated into nearly pure fractions at the outlet. The SMB process consists of several chromatographic columns which are interconnected in series to constitute a closed loop (see Figure 15.1). An effective counter-current movement of the stationary phase relative to the liquid phase is realized by periodic and simultaneous switching of the inlet and outlet ports by one column in the direction of the liquid flow. Compared to batch operation of a single chromatographic column, the SMB process offers great improvements of process performance in terms of desorbent consumption and utilization of the solid bed. In the basic SMB process all

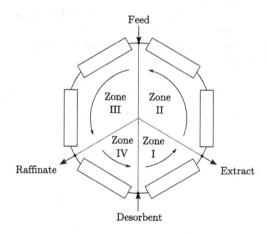

Figure 15.1: SMB configuration with six columns and four zones.

flow rates are constant and the switching of the columns is simultaneous with a fixed switching period. By introducing more degrees of freedom the efficiency of the separation can be increased further. The flow rates for instance can be varied during the switching periods (PowerFeed), the feed concentration can be varied during the switching periods (ModiCon) or asynchronous switching of the ports can be introduced (VariCol) (see Schramm et al. [144, 145]).

15.1 Mathematical modeling of adsorption processes

Accurate dynamic models of such multi-column continuous chromatographic processes consist of the dynamic process models for each single chromatographic column, the node balances which describe the connection of the columns, and the port switching. The behavior of radially homogeneous chromatographic columns is described by the General Rate Model (see Schmidt-Traub [142]).

15.1.1 General Rate Model

For both species $i = 1, 2$ the General Rate Model considers three phases, namely the instationary phase c_i which moves through the columns between the fixed bed particles, the liquid stationary phase $c_{p,i}$ inside the porous fixed bed particles, and the adsorbed stationary phase $q_{p,i}$ on the inner surface of the particles.

We assume that the columns are long and thin enough that radial concentration profiles can be neglected. The fixed bed particles are assumed to be spherical and the concentrations inside the particles are assumed to be rotationally symmetric. The governing equations in non-dimensional form are

$$\partial_t c_i = \mathrm{Pe}_i^{-1}\partial_{zz}c_i - \partial_z c_i - \mathrm{St}_i\left(c_i - c_{p,i}|_{r=1}\right), \quad (t,z) \in (0,T) \times (0,1), \quad (15.1a)$$

$$\partial_t\left((1-\varepsilon_p)q_{p,i} + \varepsilon_p c_{p,i}\right) = \eta_i\left(r^{-2}\partial_r\left(r^2\partial_r c_{p,i}\right)\right), (t,r) \in (0,T) \times (0,1), \quad (15.1b)$$

together with the boundary conditions

$$\partial_z c_i(t,0) = \mathrm{Pe}_i\left(c_i(t,0) - c^{\mathrm{in}}(t)\right), \quad \partial_r c_{p,i}(t,0) = 0, \quad (15.2a)$$

$$\partial_z c_i(t,1) = 0, \quad \partial_r c_{p,i}(t,1) = \mathrm{Bi}_i\left(c_i(t,z) - c_{p,i}(t,1)\right), \quad (15.2b)$$

with positive constants ε_p (porosity), η_i (nondimensional diffusion coefficient), Pe_i (Péclet number), St_i (Stanton number), and Bi_i (Biot number). The stationary phases are coupled by an algebraic condition, e.g., the nonlinear extended Langmuir isotherm equation

$$q_{p,i} = H_i^1 c_{p,i} + \frac{H_i^2 c_{p,i}}{1 + (k_1 c_{p,1} + k_2 c_{p,2}) c_{ref}}, \tag{15.3}$$

with non-negative constants H_i^1, H_i^2 (Henry coefficients), k_i (isotherm parameters), and reference concentration c_{ref}.

The model poses a number of difficulties:

1. The isotherm equations are algebraic constraints.
2. The time derivatives $\partial_t q_{p,i}$ and $\partial_t c_{p,i}$ are coupled on the left hand side of equation (15.1b).
3. For each point $z \in [0,1]$ in the axial direction a stationary phase equation (15.1b) is supposed to hold.
4. The stationary phase equation has a singularity for $r = 0$.

Regarding point (3), we should think of equation (15.1b) as living on the two-dimensional (z,r) domain without any derivatives in the axial direction. The coupling occurs through the boundary conditions and equation (15.1a). Gu [76] proposed to address this issue by using a low order collocation discretization of the stationary phase in each grid point of the mesh for the moving phase. We now explain this procedure in detail.

We address points (1) and (2) by elimination of $q_{p,i}$ via substitution of the algebraic constraints (15.3) into equation (15.1b). After differentiation with respect to t we obtain a system of the form

$$\begin{pmatrix} \partial_t c_{p,1} \\ \partial_t c_{p,2} \end{pmatrix} = G(c_{p,1}, c_{p,2})^{-1} \begin{pmatrix} \eta_1 \left(r^{-2} \partial_r \left(r^2 \partial_r c_{p,1} \right) \right) \\ \eta_2 \left(r^{-2} \partial_r \left(r^2 \partial_r c_{p,2} \right) \right) \end{pmatrix},$$

where the coupling 2-by-2 matrix G depends nonlinearly on $c_{p,i}$ via

$$G_{11} = (1 - \varepsilon_p) \left[H_1^1 + \frac{H_1^2}{1 + c_{ref} \sum_j k_j c_{p,j}} \left(1 - \frac{c_{ref} k_1 c_{p,1}}{1 + c_{ref} \sum_j k_j c_{p,j}} \right) \right] + \varepsilon_p,$$

$$G_{12} = (\varepsilon_p - 1) \frac{c_{ref} H_1^2 c_{p,1} k_2}{(1 + c_{ref} \sum_j k_j c_{p,j})^2},$$

$$G_{21} = (\varepsilon_p - 1) \frac{c_{ref} H_2^2 c_{p,2} k_1}{(1 + c_{ref} \sum_j k_j c_{p,j})^2},$$

$$G_{22} = (1 - \varepsilon_p) \left[H_2^1 + \frac{H_2^2}{1 + c_{ref} \sum_j k_j c_{p,j}} \left(1 - \frac{c_{ref} k_2 c_{p,2}}{1 + c_{ref} \sum_j k_j c_{p,j}} \right) \right] + \varepsilon_p.$$

This 2-by-2 matrix can be inverted with the closed formula

$$G^{-1} = \frac{1}{G_{11}G_{22} - G_{21}G_{12}} \begin{pmatrix} G_{22} & -G_{12} \\ -G_{21} & G_{11} \end{pmatrix}.$$

As proposed by Gu [76] we approximate $C_{p,i}(t,r)$ by a quadratic collocation polynomial $\varphi(r)$. We impose that $\varphi(r)$ satisfies the two boundary conditions (15.2). Thus we are left with one degree of freedom which we choose to be the point value

$$b_i(t,z) := \varphi(0.5).$$

We are lead to the form

$$\varphi(r) = 4\sigma_i(c_i(t,z) - b_i(t,z))r^2 + b_i(t,z) - \sigma_i(c_i(t,z) - b_i(t,z))$$

with the abbreviation

$$\sigma_i = \frac{Bi_i}{8 + 3Bi_i}.$$

The properties $\varphi(0.5) = b_i(t,z)$ and $\partial_r \varphi(0) = 0$ are readily verified. The second boundary condition holds true due to

$$\varphi(1) = 4\sigma_i(c_i - b_i) + b_i - \sigma_i(c_i - b_i) = 3\sigma_i(c_i - b_i) + b_i,$$
$$Bi_i(c_i - \varphi(1)) = Bi_i(c_i - 3\sigma_i(c_i - b_i) - b_i)$$
$$= \frac{Bi_i}{8 + 3Bi_i}[(8 + 3Bi_i)(c_i - b_i) - 3Bi_i(c_i - b_i)]$$
$$= 8\sigma_i(c_i - b_i) = \varphi'(1).$$

For completeness we assemble here the derivatives and surface values required for the substitution of the $c_{p,i}$ terms by b_i:

$$\frac{\partial \varphi}{\partial t}(0.5) = \frac{\partial b_i}{\partial t}, \qquad\qquad \frac{\partial \varphi}{\partial r}(r) = 8\sigma_i(c_i - b_i)r,$$
$$\frac{1}{r^2}\frac{\partial}{\partial r}\left(r^2 \frac{\partial \varphi}{\partial r}\right) = 24\sigma_i(c_i - b_i), \qquad \varphi(1) = 3\sigma_i(c_i - b_i) + b_i.$$

All in all we have transformed equations (15.1a) and (15.1b) to

$$\partial_t c_i = Pe_i^{-1}\partial_{zz}c_i - \partial_z c_i - St_i(c_i - (3\sigma_i(c_i - b_i)) + b_i), \qquad (15.4a)$$

$$\begin{pmatrix} \partial_t b_1 \\ \partial_t b_2 \end{pmatrix} = G(b_1, b_2)^{-1} \begin{pmatrix} \eta_1 24\sigma_1(c_1 - b_1) \\ \eta_2 24\sigma_2(c_2 - b_2) \end{pmatrix}, \qquad (15.4b)$$

with boundary conditions

$$\partial_z c_i(t,0) = \text{Pe}_i \left(c_i(t,0) - c^{\text{in}}(t) \right), \qquad \partial_z c_i(t,1) = 0.$$

In the case of several connected columns we use one reference flow velocity u_{ref} for the non-dimensionalization. For a flow velocity $u_j \neq u_{\text{ref}}$ in zone $j = \text{I}, \ldots, \text{IV}$ we have to multiply the right hand sides of equations (15.1) or (15.4), respectively, with the quotient u_j/u_{ref}.

15.1.2 Mass balances

The model for the whole SMB process consists of a fixed number N_{col} of columns described by the General Rate Model and mass balances at the ports between the columns. The concentrations of column j are denoted by a superscript j. In the ModiCon variant, the process is controlled by the time-independent flow rates Q_{De} (desorbent), Q_{Ex} (extract), Q_{Rec} (recycle), Q_{Fe} (feed), and the time-dependent feed concentration $c_{\text{Fe}}(t)$. The remaining flow rates, which are the raffinate flow rate Q_{Ra} and the zone flow rates $Q_I, \ldots, Q_{IV}$, are fully determined by conservation of mass via

$$Q_{\text{Ra}} = Q_{\text{De}} - Q_{\text{Ex}} + Q_{\text{Fe}},$$
$$Q_I = Q_{\text{De}} + Q_{\text{Rec}}, \qquad\qquad Q_{II} = Q_I - Q_{\text{Ex}},$$
$$Q_{III} = Q_{II} + Q_{\text{Fe}}, \qquad\qquad Q_{IV} = Q_{III} - Q_{\text{Ra}} = Q_{\text{Rec}}.$$

The inflow concentrations of each column are the outflow concentrations of the preceding column, except for the column after the feed and after the desorbent ports which can be calculated from the feed concentration $c_{\text{Fe},i}$ and from the outflow concentrations $c_{\cdot,i}^{\text{out}}$ of the previous column according to

$$c_{I,i}^{\text{in}} Q_I = c_{IV,i}^{\text{out}} Q_{IV}, \qquad\qquad c_{III,i}^{\text{in}} Q_{III} = c_{II,i}^{\text{out}} Q_{II} + c_{\text{Fe},i} Q_{\text{Fe}},$$

for $i = 1, 2$. With the port concentrations and the flow rates the feed, extract, and raffinate masses, and the product purities can be calculated via

$$m_{\text{Fe},i}(t) = \int_0^t c_{\text{Fe},i}(\tau) Q_{\text{Fe}} d\tau, \qquad\qquad m_{\text{Ex},i}(t) = \int_0^t c_{I,i}^{\text{out}}(\tau,1) Q_{\text{Ex}} d\tau,$$

$$m_{\text{Ra},i}(t) = \int_0^t c_{III,i}^{\text{out}}(\tau,1) Q_{\text{Ra}} d\tau,$$

$$\text{Pur}_{\text{Ex}}(t) = \frac{m_{\text{Ex},1}(t)}{m_{\text{Ex},1}(t) + m_{\text{Ex},2}(t)}, \qquad\qquad \text{Pur}_{\text{Ra}}(t) = \frac{m_{\text{Ra},2}(t)}{m_{\text{Ra},1}(t) + m_{\text{Ra},2}(t)}.$$

15.1.3 Objective and constraints

We consider the optimization of an SMB process with variable feed concentration (ModiCon process) which minimizes desorbent consumption

$$\int_0^T Q_{De}(t)\,dt$$

subject to purity constraints for the two product streams

$$\text{Pur}_{Ex}(T) \geq \text{Pur}_{min} \quad \text{and} \quad \text{Pur}_{Ra}(T) \geq \text{Pur}_{min}$$

at a constant feed flow Q_{Fe} but varying feed concentration $c_{Fe}(t)$. Over one period T the average feed concentration must be equal to the given feed concentration c_{Fe}^{SMB} of a reference SMB process.

At the end of each period the switching of ports leads to a generalized periodicity constraint of the form

$$\left.\begin{aligned} c_i^j(0,.) - c_i^{\text{succ}(j)}(T,.) &= 0, \\ b_i^j(0,.) - b_i^{\text{succ}(j)}(T,.) &= 0, \end{aligned}\right\} \quad i = 1,2, j = 1,\ldots,N_{col},$$

where $\text{succ}(j)$ denotes the index of the column which is the successor of column j in the investigated SMB configuration.

Furthermore we require the total feed mass of one period to satisfy

$$m_{Fe}(T) = c_{Fe}^{SMB} Q_{Fe} T, \tag{15.5}$$

where c_{Fe}^{SMB} is a given feed concentration of a (non-ModiCon) SMB reference process.

The remaining constraints bound the maximum and minimum feed concentration

$$c_{Fe,max} \geq c_{Fe}(t) \geq 0$$

and the flow rates

$$Q_{max} \geq Q_{De}, Q_{Ex}, Q_{Fe}, Q_{Ra}, Q_{Re}, Q_I, Q_{II}, Q_{III}, Q_{IV} \geq Q_{min}.$$

15.2 Numerical results

The results in this chapter were computed for EMD–53986 enantiomer separation. EMD–53986, or 5-(1,2,3,4-tetra-hydroquinolin-6-yl)-6-methyl-3,6-dihydro-1,3,4-thiadiazin-2-one (see Figure 15.2), is a chiral precursor for a pharmaceutical

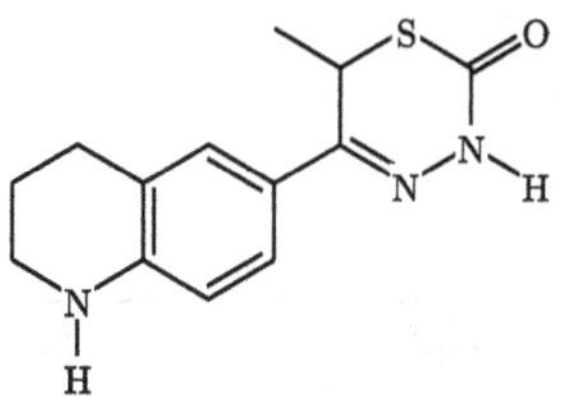

Figure 15.2: Chemical structure of EMD–53986.

reagent (see, e.g., Jupke [91] as cited by Küpper [100]). Only the R-enantiomer has pharmaceutical activity and needs to be separated from the S-enantiomer after chemical synthesis. We list the model parameters (taken from Küpper [100]) in Table 15.1. Further model quantities are derived from these parameters which we display in Table 15.2.

We computed the solution with $n_{MS} = 24$ shooting intervals on a two level hierarchy of spatial FDM grids with 21 and 81 equidistant grid points for each of the $N_{col} = 6$ columns and each species. The relative accuracy for the time-stepping scheme was set to 10^{-5} on the coarse and 10^{-6} on the fine level and the GINKO termination tolerance was $5 \cdot 10^{-3}$.

The optimization problem becomes more and more difficult for higher values of product purity Pur_{min}. We had to successively generate primal starting values via a naive homotopy approach using ascending values for $Pur_{min} = 0.8, 0.9, 0.93, 0.95$ on the coarse level. For $Pur_{min} = 0.95$, GINKO needed 9 iterations on the coarse level and then 11 iterations on the fine level with coarse grid derivatives. The computed κ-estimates suggest $[\hat{\kappa}] \leq 0.66$. For the last four iterations only two LISA are needed for each inexact solution of the linearized systems. Table 15.3 shows the optimal values for the ModiCon SMB separation of EMB–53986 enantiomers. We display the optimal feed concentration profile in Figure 15.4 and the optimal moving concentration fronts of the moving phase for one period in Figure 15.3. We can observe that the two concentration profiles travel to the right with different velocities and thus there is almost only slow substance present at the extract port after column 1 and almost only fast substance present at the raffinate port after column 5.

The solution was computed on four cores within a total wall time of 98 min. Due to memory restrictions for the IND tape sizes, finer grid levels were not possible. In the solution there are between 24 and 177 integration steps per shooting interval with an average of 42.1 steps per interval.

Symbol	Value	Unit	Description
L	9.0	cm	column length
D	2.5	cm	column diameter
ε_p	0.567	–	particle void fraction
ε_b	0.353	–	bulk void fraction
d_p	0.002	cm	particle diameter
ρ	0.799	g/cm^3	fluid density
ν	0.012	$g/(cm\,s)$	fluid viscosity
D_p	0.001	cm^2/s	particle diffusion (estimated)
$k_{app,1}$	1.5e-4	$1/s$	apparent mass transfer coefficient
$k_{app,2}$	2.0e-4	$1/s$	apparent mass transfer coefficient
H_1^1	2.054	–	Henry coefficient
H_2^1	2.054	–	Henry coefficient
H_1^2	19.902	–	Henry coefficient
H_2^2	5.847	–	Henry coefficient
k_1	472.0	cm^3/g	isotherm parameter
k_2	129.0	cm^3/g	isotherm parameter
c_{ref}	2.5e-3	g/cm^3	reference concentration
Pur_{min}	95	%	minimum product purity
c_{Fe}^{SMB}	2.5e-3	g/cm^3	reference feed concentration
$c_{Fe,max}$	1.25e-2	g/cm^3	maximum feed concentration
Q_{max}	300	ml/min	maximum flow rate
Q_{min}	30	ml/min	minimum flow rate

Table 15.1: Model and optimization parameters for the ModiCon SMB process to separate EMD–53986 enantiomers.

Symbol	Formula	Description
$k_{\text{eff},i}$	$\dfrac{6}{d_p}k_{\text{app},i}$	effective mass transfer coefficient
$k_{1,i}$	$\dfrac{d_p}{6}\dfrac{k_{\text{eff},i}15\varepsilon_p D_p}{15\varepsilon_p D_p - (d_p/2)^2 k_{\text{eff},i}}$	mass transfer coefficient
Bi_i	$\dfrac{k_{1,i}d_p}{2\varepsilon_p D_p}$	Biot number
u_{ref}	$\dfrac{4Q_{\text{III}}}{\pi D^2 \varepsilon_b}$	reference flow velocity
u_j	$\dfrac{4Q_j}{\pi D^2 \varepsilon_b}$	flow velocity
Re_j	$\dfrac{\rho u_j d_p}{\nu}$	Reynolds number
Pe_j	$\dfrac{0.2+0.011(\text{Re}_j\varepsilon_b)^{0.48}}{\varepsilon_b}\dfrac{L}{d_p}$	Péclet number
η_j	$\dfrac{4\varepsilon_p D_p L}{d_p^2 u_j}$	particle diffusion coefficient
St_i^j	$3\text{Bi}_i\eta_j\dfrac{1-\varepsilon_b}{\varepsilon_b}$	Stanton number

Table 15.2: Derived model quantities for the ModiCon SMB process. Index $i = 1,2$ denotes the species, index $j = \text{I},\ldots,\text{IV}$ the zone.

Description	Symbol	Optimal value	Unit
period duration	T	10.68	min
desorbent flow	Q_{De}	86.33	ml/min
extract flow	Q_{Ex}	86.33	ml/min
feed flow	Q_{Fe}	30.00	ml/min
recycle flow	Q_{Re}	30.00	ml/min
objective	$\int_0^T Q_{\text{De}}(t)\mathrm{d}t$	921.8	ml

Table 15.3: Optimal values for the ModiCon SMB separation of EMB–53986 enantiomers.

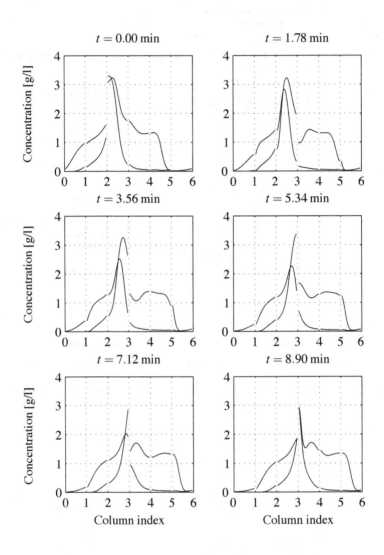

Figure 15.3: Traveling concentration profiles over one period $t \in [0, T]$. The six columns are arranged from left to right in each panel. The feed port is located after column 3, extract after column 1, raffinate after colum 5 and desorbent after column 6.

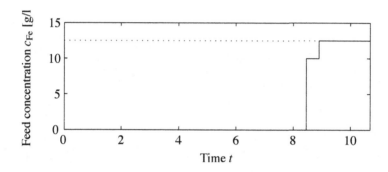

Figure 15.4: Optimal feed concentration profile c_{Fe} for ModiCon SMB separation of EMB–53986 enantiomers. All feed mass is injected at the end of the period with maximum concentration $c_{Fe,max}$, subject to satisfaction of total feed mass constraint (15.5).

16 Conclusions and future work

In this thesis we have developed a numerical method based on Direct Multiple Shooting for OCPs with time-periodic PDE constraints. We have achieved an asymptotically optimal scale-up of the numerical effort with the number of spatial discretization points based on inexact SQP with an inner generalized Newton-Picard preconditioned LISA which features extensive structure exploitation in a two-stage solution process for the possibly nonconvex QPs for which we have developed a condensing approach and a PASM. We have implemented a numerical code called MUSCOP and have demonstrated the applicability, efficiency, and reliability of the proposed methods on PDE OCPs from illustrating academic to challenging real-world applications.

Our research inspires a number of now exposed questions for future research directions and projects. We want to conclude this thesis with a list of the most obvious ones:

Convergence analysis for PASMs for nonconvex QPs. We have developed numerical techniques for the solution of nonconvex QPs with PASMs. In our experience the resulting QP solver works unexpectedly well on these difficult problems. We believe it worthwhile to construct examples for which it does not work or investigate proofs if none can be found. These examples, if found, might also serve as the basis for further improvement of the numerical method.

NMT for inexact SQP. We have presented a NMT globalization strategy for the class of LISA-Newton based inexact SQP methods. However, no proof of convergence exists for this approach. We conjecture such a proof is possible even for the inexact SQP case on the basis of RMT techniques.

A-posteriori mesh error estimation and mesh refinement. We have computed all numerical examples in this thesis on uniformly refined spatial meshes. Obviously locally refined meshes promise a great improvement of the ratio of numerical effort vs. accuracy (see, e.g., Becker and Rannacher [14], Meidner and Vexler [113], Hesse [83]). Furthermore, we should refine the fine and coarse grid according to two different goals: The fine grid for highest accuracy (in terms of the question which inspires the problem) and the coarse grid for best contraction, i.e., smallest κ. Moreover the required global error estimators should be exploited to trigger fine grid refinement as soon as the inexact Simplified Newton increment becomes smaller than the grid interpolation error.

DAESOL-II tape management. The tape management of DAESOL-II is currently the memory bottleneck of MUSCOP. We propose to implement asynchronous swapping of tape entries in and out of main memory to hard disk. Because the tape must always be read in sequential order either forward or backward, it is possible to prompt swapping of the required blocks into main memory in advance. This process does not require CPU cycles on most current hardware platforms and can be performed concurrently with the remaining required computations.

Load balancing techniques. For simplicity we have only implemented regular distribution of Multiple Shooting IVPs for simulation and IND to worker processes in parallel. As we have demonstrated in Chapter 13 even a simple adaptive greedy heuristic could considerably improve the speed-up of MUSCOP.

Computations on a cluster computer. The numerical results which we have presented in this thesis were computed on a desktop machine on four CPU cores. We have designed MUSCOP to also run on a distributed cluster computer. This approach would also mitigate the memory bottleneck issue because memory consumption is also parallelized in the proposed algorithm on the Multiple Shooting structure.

Nonlinear instationary 2D and 3D problems. As we can see from the analysis (e.g., Theorem 6.7), the proposed methods are not generally restricted by the dimensionality of the considered geometry. After completion of the projects mentioned in the three previous paragraphs we believe it is possible to treat even larger instationary problems in 2D. For 3D problems we anticipate that numerical fill-in in the direct sparse linear algebra routines within DAESOL-II can become a bottleneck and would have to be addressed before.

Bibliography

[1] A. Agarwal, L.T. Biegler, and S.E. Zitney. Simulation and optimization of pressure swing adsorption systems using reduced-order modeling. *Industrial & Engineering Chemistry Research*, 48(5):2327–2343, 2009.

[2] J. Albersmeyer. *Adjoint based algorithms and numerical methods for sensitivity generation and optimization of large scale dynamic systems.* PhD thesis, Ruprecht–Karls–Universität Heidelberg, 2010. URL http://www.ub.uni-heidelberg.de/archiv/11651/.

[3] J. Albersmeyer and H.G. Bock. Efficient sensitivity generation for large scale dynamic systems. Technical report, SPP 1253 Preprints, University of Erlangen, 2009.

[4] J. Albersmeyer and M. Diehl. The Lifted Newton method and its application in optimization. *SIAM Journal on Optimization*, 20(3):1655–1684, 2010.

[5] J. Albersmeyer and C. Kirches. The SolvIND webpage, 2007. URL http://www.solvind.org.

[6] M. Arioli, I.S. Duff, N.I.M. Gould, J.D. Hogg, J.A. Scott, and H.S. Thorne. The HSL mathematical software library, 2007.

[7] U. Ascher and M.R. Osborne. A note on solving nonlinear equations and the natural criterion function. *Journal of Optimization Theory and Applications*, 55(1):147–152, 1987.

[8] V. Bär. Ein Kollokationsverfahren zur numerischen Lösung allgemeiner Mehrpunktrandwertaufgaben mit Schalt- und Sprungbedingungen mit Anwendungen in der optimalen Steuerung und der Parameteridentifizierung. Diploma thesis, Rheinische Friedrich–Wilhelms–Universität zu Bonn, 1983.

[9] A. Battermann and M. Heinkenschloss. Preconditioners for Karush-Kuhn-Tucker matrices arising in the optimal control of distributed systems. In

Control and estimation of distributed parameter systems (Vorau, 1996), volume 126 of *Internat. Ser. Numer. Math.*, pages 15–32. Birkhäuser, Basel, 1998.

[10] A. Battermann and E.W. Sachs. Block preconditioners for KKT systems in PDE-governed optimal control problems. In *Fast solution of discretized optimization problems (Berlin, 2000)*, volume 138 of *Internat. Ser. Numer. Math.*, pages 1–18. Birkhäuser, Basel, 2001.

[11] I. Bauer. *Numerische Verfahren zur Lösung von Anfangswertaufgaben und zur Generierung von ersten und zweiten Ableitungen mit Anwendungen bei Optimierungsaufgaben in Chemie und Verfahrenstechnik*. PhD thesis, Universität Heidelberg, 1999. URL http://www.ub.uni-heidelberg.de/archiv/1513.

[12] I. Bauer, H.G. Bock, S. Körkel, and J.P. Schlöder. Numerical methods for initial value problems and derivative generation for DAE models with application to optimum experimental design of chemical processes. In *Scientific Computing in Chemical Engineering II*, pages 282–289. Springer, 1999.

[13] I. Bauer, H.G. Bock, and J.P. Schlöder. DAESOL – a BDF-code for the numerical solution of differential algebraic equations, 1999.

[14] R. Becker and R. Rannacher. An optimal control approach to error estimation and mesh adaptation in finite element methods. *Acta Numerica 2000*, pages 1–101, 2001.

[15] R.E. Bellman. *Dynamic Programming*. University Press, Princeton, N.J., 6th edition, 1957.

[16] M. Benzi, G.H. Golub, and J. Liesen. Numerical solution of saddle–point problems. *Acta Numerica*, 14:1–137, 2005.

[17] M.J. Best. An algorithm for the solution of the parametric quadratic programming problem. In H. Fischer, B. Riedmüller, and S. Schäffler, editors, *Applied Mathematics and Parallel Computing – Festschrift for Klaus Ritter*, chapter 3, pages 57–76. Physica-Verlag, Heidelberg, 1996.

[18] L.T. Biegler. Solution of dynamic optimization problems by successive quadratic programming and orthogonal collocation. *Computers and Chemical Engineering*, 8:243–248, 1984.

[19] G. Biros and O. Ghattas. Parallel Lagrange-Newton-Krylov-Schur methods for PDE-constrained optimization. Part I: The Krylov-Schur solver. *SIAM Journal on Scientific Computing*, 27(2):687–713, 2005.

[20] G. Biros and O. Ghattas. Parallel Lagrange-Newton-Krylov-Schur methods for PDE-constrained optimization. Part II: The Lagrange-Newton solver and its application to optimal control of steady viscous flows. *SIAM Journal on Scientific Computing*, 27(2):714–739, 2005.

[21] J. Blue, P. Fox, W. Fullerton, D. Gay, E. Grosse, A. Hall, L. Kaufman, W. Petersen, and N. Schryer. PORT mathematical subroutine library, 1997. URL http://www.bell-labs.com/project/PORT/.

[22] H.G. Bock. Numerical treatment of inverse problems in chemical reaction kinetics. In K.H. Ebert, P. Deuflhard, and W. Jäger, editors, *Modelling of Chemical Reaction Systems*, volume 18 of *Springer Series in Chemical Physics*, pages 102–125. Springer, Heidelberg, 1981. URL http://www.iwr.uni-heidelberg.de/groups/agbock/FILES/Bock1981.pdf.

[23] H.G. Bock. Recent advances in parameter identification techniques for ODE. In P. Deuflhard and E. Hairer, editors, *Numerical Treatment of Inverse Problems in Differential and Integral Equations*, pages 95–121. Birkhäuser, Boston, 1983. URL http://www.iwr.uni-heidelberg.de/groups/agbock/FILES/Bock1983.pdf.

[24] H.G. Bock. *Randwertproblemmethoden zur Parameteridentifizierung in Systemen nichtlinearer Differentialgleichungen*, volume 183 of *Bonner Mathematische Schriften*. Universität Bonn, Bonn, 1987. URL http://www.iwr.uni-heidelberg.de/groups/agbock/FILES/Bock1987.pdf.

[25] H.G. Bock and K.J. Plitt. A Multiple Shooting algorithm for direct solution of optimal control problems. In *Proceedings of the 9th IFAC World Congress*, pages 242–247, Budapest, 1984. Pergamon Press. URL http://www.iwr.uni-heidelberg.de/groups/agbock/FILES/Bock1984.pdf.

[26] H.G. Bock, E.A. Kostina, and J.P. Schlöder. On the role of natural level functions to achieve global convergence for damped Newton methods. In M.J.D. Powell and S. Scholtes, editors, *System Modelling and Optimization. Methods, Theory and Applications*, pages 51–74. Kluwer, 2000.

[27] H.G. Bock, W. Egartner, W. Kappis, and V. Schulz. Practical shape optimization for turbine and compressor blades by the use of PRSQP methods. *Optimization and Engineering*, 3(4):395–414, 2002.

[28] H.G. Bock, A. Potschka, S. Sager, and J.P. Schlöder. On the connection between forward and optimization problem in one-shot one-step methods. In G. Leugering, S. Engell, A. Griewank, M. Hinze, R. Rannacher, V. Schulz, M. Ulbrich, and S. Ulbrich, editors, *Constrained Optimization and Optimal Control for Partial Differential Equations*, volume 160 of *International Series of Numerical Mathematics*, pages 37–49. Springer Basel, 2011.

[29] Dietrich Braess. *Finite elements*. Cambridge University Press, Cambridge, 3rd edition, 2007.

[30] J.H. Bramble and J.E. Pasciak. A preconditioning technique for indefinite systems resulting from mixed approximations of elliptic problems. *Mathematics of Computation*, 50(181):1–17, 1988.

[31] A.N. Brooks and T.J.R. Hughes. Streamline upwind/Petrov-Galerkin formulations for convection dominated flows with particular emphasis on the incompressible Navier-Stokes equations. *Computational Methods in Applied Mechanics and Engineering*, 32(1-3):199–259, 1982.

[32] K. Burrage, J. Erhel, B. Pohl, and A. Williams. A deflation technique for linear systems of equations. *SIAM Journal on Scientific Computing*, 19(4): 1245–1260, 1998.

[33] R.H. Byrd, F.E. Curtis, and J. Nocedal. An inexact SQP method for equality constrained optimization. *SIAM Journal on Optimization*, 19(1):351–369, 2008.

[34] A.R. Conn, K. Scheinberg, and L.N. Vicente. *Introduction to derivative-free optimization*, volume 8 of *MPS/SIAM Series on Optimization*. Society for Industrial and Applied Mathematics (SIAM), Philadelphia, PA, 2009.

[35] J. Dahl and L. Vandenberghe. CVXOPT user's guide, release 1.1.3, 2010. URL http://abel.ee.ucla.edu/cvxopt/userguide/index.html.

[36] G.B. Dantzig. *Linear Programming and Extensions*. Princeton University Press, 1963.

[37] R. Dautray and J.-L. Lions. Evolution problems I. In A. Craig, editor, *Mathematical analysis and numerical methods for science and technology*, volume 5. Springer, 1992.

[38] D.F. Davidenko. On a new method of numerical solution of systems of nonlinear equations. *Doklady Akademii nauk SSSR*, 88:601–602, 1953.

[39] T.A. Davis. Algorithm 832: UMFPACK – an unsymmetric-pattern multifrontal method with a column pre-ordering strategy. *ACM Transactions on Mathematical Software*, 30:196–199, 2004.

[40] V. de la Torre, A. Walther, and L.T. Biegler. Optimization of periodic adsorption processes with a novel problem formulation and nonlinear programming algorithm. In *AD 2004 – Fourth International Workshop on Automatic Differentiation, July 19-23, 2004, Argonne National Laboratory, USA*, 2004. (extended conference abstract).

[41] R.S. Dembo, S.C. Eisenstat, and T. Steihaug. Inexact Newton methods. *SIAM Journal on Numerical Analysis*, 19(2):400–408, 1982.

[42] P. Deuflhard. *Ein Newton-Verfahren bei fastsingulärer Funktionalmatrix zur Lösung von nichtlinearen Randwertaufgaben mit der Mehrzielmethode*. PhD thesis, Universität zu Köln, 1972.

[43] P. Deuflhard. A Modified Newton Method for the Solution of Ill-conditioned Systems of Nonlinear Equations with Applications to Multiple Shooting. *Numerische Mathematik*, 22:289–311, 1974.

[44] P. Deuflhard. *Newton Methods for Nonlinear Problems. Affine Invariance and Adaptive Algorithms*, volume 35 of *Springer Series in Computational Mathematics*. Springer, 2006.

[45] P. Deuflhard, R. Freund, and A. Walter. Fast secant methods for the iterative solution of large nonsymmetric linear systems. *IMPACT of Computing in Science and Engineering*, 2(3):244–276, 1990.

[46] L. Di Gaspero. QuadProg++, 2010. URL http://www.diegm.uniud.it/digaspero/index.php/software/.

[47] M. Diehl. *Real-Time Optimization for Large Scale Nonlinear Processes*. PhD thesis, Universität Heidelberg, 2001. URL http://www.ub.uni-heidelberg.de/archiv/1659/.

[48] M. Diehl, A. Walther, H.G. Bock, and E. Kostina. An adjoint-based SQP algorithm with quasi-Newton Jacobian updates for inequality constrained optimization. *Optimization Methods and Software*, 2009.

[49] J. Dongarra, V. Eijkhout, and A. Kalhan. LAPACK working note 99: Reverse communication interface for linear algebra templates for iterative methods. Technical Report UT-CS-95-292, University of Tennessee, 1995.

[50] N. Dunford and J.T. Schwartz. Linear operators part I: General theory. In R. Courant, L. Bers, and J.J. Stoker, editors, *Pure and applied mathematics*, volume VII. Wiley, New York, 1958.

[51] J. Fernández, M. Anguita, E. Ros, and J.L. Bernier. SCE toolboxes for the development of high–level parallel applications. In *Proceedings of the 6th International Conference Computational Science – ICCS 2006, Reading, United Kingdom, Part II*, volume 3992, pages 518–525, 2006.

[52] H.J. Ferreau. An online active set strategy for fast solution of parametric quadratic programs with applications to predictive engine control. Diploma thesis, Ruprecht–Karls–Universität Heidelberg, 2006. URL http://homes.esat.kuleuven.be/~jferreau/pdf/thesisONLINE.pdf.

[53] H.J. Ferreau, H.G. Bock, and M. Diehl. An online active set strategy to overcome the limitations of explicit MPC. *International Journal of Robust and Nonlinear Control*, 18(8):816–830, 2008.

[54] FICO. *FICO(TM) Xpress Optimization Suite, Xpress-Optimizer Reference manual, Release 20.00*. Fair Isaac Corporation, Warwickshire, UK, 2009.

[55] R. Fletcher. Resolving degeneracy in quadratic programming. *Annals of Operations Research*, 46-47:307–334, 1993.

[56] C.A. Floudas and P.M. Pardalos, editors. *State of the art in global optimization: computational methods and applications*. Springer, Dordrecht, 1995.

[57] H. Gajewski, K. Gröger, and K. Zacharias. *Nichtlineare Operatorgleichungen und Operatordifferentialgleichungen*. Akademie-Verlag, Berlin, 1974.

[58] E.M. Gertz and S.J. Wright. Object-oriented software for quadratic programming. *ACM Transactions on Mathematical Software*, 29:58–81, 2003.

[59] P.E. Gill, W. Murray, M.A. Saunders, and M.H. Wright. Procedures for optimization problems with a mixture of bounds and general linear constraints. *ACM Transactions on Mathematical Software*, 10(3):282–298, 1984.

[60] P.E. Gill, W. Murray, and M.A. Saunders. *User's Guide For QPOPT 1.0: A Fortran Package For Quadratic Programming*, 1995. URL http://www.sbsi-sol-optimize.com/manuals/QPOPT%20Manual.pdf.

[61] G.H. Golub and C.F. van Loan. *Matrix Computations*. Johns Hopkins University Press, Baltimore, 3rd edition, 1996.

[62] J. Gondzio. HOPDM (version 2.12) – a fast LP solver based on a primal-dual interior point method. *European Journal of Operational Research*, 85 (1):221–225, 1995.

[63] N.I.M. Gould. On practical conditions for the existence and uniqueness of solutions to the general equality quadratic programming problem. *Mathematical Programming*, 32(1):90–99, 1985.

[64] N.I.M. Gould. An algorithm for large-scale quadratic programming. *IMA Journal of Numerical Analysis*, 11(3):299–324, 1991.

[65] N.I.M. Gould and Ph.L. Toint. Nonlinear programming without a penalty function or a filter. *Mathematical Programming, Series A*, 122:155–196, 2010.

[66] N.I.M. Gould and P.L. Toint. A quadratic programming bibliography. Technical Report 2000-1, Rutherford Appleton Laboratory, Computational Science and Engineering Department, 2010.

[67] N.I.M. Gould, M.E. Hribar, and J. Nocedal. On the solution of equality constrained quadratic programming problems arising in optimization. *SIAM Journal on Scientific Computing*, 23(4):1376–1395, 2001.

[68] N.I.M. Gould, D. Orban, and P.L. Toint. CUTEr testing environment for optimization and linear algebra solvers, 2002. URL http://cuter.rl.ac.uk/cuter-www/.

[69] N.I.M. Gould, D. Orban, and Ph.L. Toint. GALAHAD, a library of thread-safe Fortran 90 packages for large-scale nonlinear optimization. *ACM Transactions on Mathematical Software*, 29(4):353–372, 2004.

[70] A. Griewank. *Evaluating Derivatives, Principles and Techniques of Algorithmic Differentiation*. Number 19 in Frontiers in Applied Mathematics. SIAM, Philadelphia, 2000.

[71] A. Griewank. Projected Hessians for preconditioning in One-Step One-Shot design optimization. In *Large-Scale Nonlinear Optimization*, volume 83 of *Nonconvex Optimization and Its Applications*, pages 151–171. Springer US, 2006.

[72] A. Griewank, D. Juedes, and J. Utke. Algorithm 755: ADOL-C: A package for the automatic differentiation of algorithms written in C/C++. *ACM Transactions on Mathematical Software*, 22(2):131–167, 1996.

[73] A. Griewank, D. Juedes, H. Mitev, J. Utke, O. Vogel, and A. Walther. ADOL-C: A package for the automatic differentiation of algorithms written in C/C++. Technical report, Technical University of Dresden, Institute of Scientific Computing and Institute of Geometry, 1999. Updated version of the paper published in *ACM Trans. Math. Software* 22:131–167, 1996.

[74] A. Griewank, J. Utke, and A. Walther. Evaluating higher derivative tensors by forward propagation of univariate Taylor series. *Mathematics of Computation*, pages 1117–1130, 2000.

[75] W. Gropp, E. Lusk, and A. Skjellum. *Using MPI: Portable Parallel Programming with the Message Passing Interface*. Scientific and Engineering Computation Series. MIT Press, Cambridge, MA, USA, 2nd edition, 1999.

[76] T. Gu. *Mathematical Modelling and Scale Up of Liquid Chromatography*. Springer Verlag, New York, 1995.

[77] W. Hackbusch. Fast numerical solution of time-periodic parabolic problems by a multigrid method. *SIAM Journal on Scientific and Statistical Computing*, 2(2):198–206, 1981.

[78] E. Hairer, S.P. Nørsett, and G. Wanner. *Solving Ordinary Differential Equations I*, volume 8 of *Springer Series in Computational Mathematics*. Springer, Berlin, 2nd edition, 1993.

[79] P. Hartman. *Ordinary differential equations*, volume 38 of *Classics in Applied Mathematics*. SIAM, second edition, 2002.

[80] S.B. Hazra, V. Schulz, J. Brezillon, and N.R. Gauger. Aerodynamic shape optimization using simultaneous pseudo-timestepping. *Journal of Computational Physics*, 204(1):46–64, 2005.

[81] M. Heinkenschloss and D. Ridzal. An Inexact Trust-Region SQP method with applications to PDE-constrained optimization. In K. Kunisch, G. Of, and O. Steinbach, editors, *Proceedings of ENUMATH 2007, the 7th European Conference on Numerical Mathematics and Advanced Applications, Graz, Austria, September 2007*. Springer Berlin Heidelberg, 2008.

[82] M. Heinkenschloss and L.N. Vicente. Analysis of Inexact Trust-Region SQP algorithms. *SIAM Journal on Optimization*, 12(2):283–302, 2001.

[83] H.K. Hesse. *Multiple Shooting and Mesh Adaptation for PDE Constrained Optimization Problems*. PhD thesis, University of Heidelberg, 2008.

[84] J.S. Hesthaven and T. Warburton. *Nodal Discontinuous Galerkin Methods*, volume 54 of *Texts in Applied Mathematics*. Springer New York, 2008.

[85] J.S. Hesthaven, S. Gottlieb, and D. Gottlieb. *Spectral methods for time-dependent problems*, volume 21 of *Cambridge Monographs on Applied and Computational Mathematics*. Cambridge University Press, Cambridge, 2007.

[86] M. Hinze, R. Pinnau, M. Ulbrich, and S. Ulbrich. *Optimization with PDE Constraints*. Springer, New York, 2009.

[87] D. Horstmann. From 1970 until present: The Keller-Segel model in chemotaxis and its consequences I. *Jahresbericht der DMV*, 105(3):103–165, 2003.

[88] D. Horstmann. From 1970 until present: The Keller-Segel model in chemotaxis and its consequences II. *Jahresbericht der DMV*, 106(2):51–69, 2004.

[89] IBM ILOG. *IBM ILOG CPLEX V12.1, User's Manual for CPLEX*. IBM Corp., New York, USA, 2009.

[90] H. Jarausch and W. Mackens. Numerical treatment of bifurcation branches by adaptive condensation. In *Numerical methods for bifurcation problems (Dortmund, 1983)*, volume 70 of *Internat. Schriftenreihe Numer. Math.*, pages 296–309. Birkhäuser, Basel, 1984.

[91] A. Jupke. *Experimentelle Modellvalidierung und modellbasierte Auslegung von Simulated Moving Bed (SMB) Chromatographieverfahren*, volume 807 of *Fortschrittberichte VDI, Reihe 3*. VDI-Verlag, Düsseldorf, 2004. Dissertation, Universität Dortmund.

[92] W. Karush. Minima of functions of several variables with inequalities as side conditions. Master's thesis, Department of Mathematics, University of Chicago, 1939.

[93] Y. Kawajiri and L.T. Biegler. Optimization strategies for Simulated Moving Bed and PowerFeed processes. *AIChE Journal*, 52(4):1343–1350, 2006.

[94] Y. Kawajiri and L.T. Biegler. Large scale nonlinear optimization for asymmetric operation and design of simulated moving beds. *Journal of Chromatography A*, 1133:226–240, 2006.

[95] E.F. Keller and L.A. Segel. Model for chemotaxis. *Journal of Theoretical Biology*, 30(2):225–234, 1971.

[96] C. Kirches, H.G. Bock, J.P. Schlöder, and S. Sager. A factorization with update procedures for a KKT matrix arising in direct optimal control. *Mathematical Programming Computation*, 2010. URL http://www.optimization-online.org/DB_HTML/2009/11/2456.html. Submitted.

[97] C. Kirches, H.G. Bock, J.P. Schlöder, and S. Sager. Block structured quadratic programming for the direct multiple shooting method for optimal control. *Optimization Methods and Software*, 26(2):239–257, 2011.

[98] E. Kostina. The long step rule in the bounded-variable dual simplex method: Numerical experiments. *Mathematical Methods of Operations Research*, 55 (3):413–429, 2002.

[99] H.W. Kuhn and A.W. Tucker. Nonlinear programming. In J. Neyman, editor, *Proceedings of the Second Berkeley Symposium on Mathematical Statistics and Probability*, pages 481–492, Berkeley, 1951. University of California Press.

[100] A. Küpper. *Optimization, State Estimation, and Model Predictive Control of Simulated Moving Bed Processes*, volume 2010,1 of *Schriftenreihe des Lehrstuhls für Systemdynamik und Prozessführung*. Prof. Dr.-Ing. Sebastian Engell, Dortmund, 2010. Dissertation.

[101] S. Lauer. SQP–Methoden zur Behandlung von Problemen mit indefiniter reduzierter Hesse–Matrix. Diploma thesis, Ruprecht–Karls–Universität Heidelberg, 2010.

[102] D. Lebiedz and U. Brandt-Pollmann. Manipulation of Self-Aggregation Patterns and Waves in a Reaction-Diffusion System by Optimal Boundary Control Strategies. *Physical Review Letters*, 91(20):208301, 2003.

[103] R.B. Lehoucq and D.C. Sorensen. Deflation techniques for an implicitly restarted Arnoldi iteration. *SIAM Journal on Matrix Analysis and Applications*, 17(4):789–821, 1996.

[104] R.B. Lehoucq, D.C. Sorensen, and C. Yang. *ARPACK Users' Guide: Solution of Large–Scale Eigenvalue Problems with Implicitly Restarted Arnoldi Methods*. Society for Industrial and Applied Mathematics (SIAM), 1998.

[105] D.B. Leineweber. The theory of MUSCOD in a nutshell. IWR-Preprint 96-19, Universität Heidelberg, 1996.

[106] D.B. Leineweber. *Efficient reduced SQP methods for the optimization of chemical processes described by large sparse DAE models*, volume 613 of *Fortschritt-Berichte VDI Reihe 3, Verfahrenstechnik*. VDI Verlag, Düsseldorf, 1999.

[107] D.B. Leineweber, I. Bauer, A.A.S. Schäfer, H.G. Bock, and J.P. Schlöder. An efficient multiple shooting based reduced SQP strategy for large-scale dynamic process optimization (Parts I and II). *Computers and Chemical Engineering*, 27:157–174, 2003.

[108] R.J. LeVeque. *Finite volume methods for hyperbolic problems*. Cambridge Texts in Applied Mathematics. Cambridge University Press, Cambridge, 2002.

[109] R.J. LeVeque. *Finite difference methods for ordinary and partial differential equations*. Society for Industrial and Applied Mathematics (SIAM), Philadelphia, PA, 2007.

[110] K. Lust, D. Roose, A. Spence, and A. R. Champneys. An adaptive Newton-Picard algorithm with subspace iteration for computing periodic solutions. *SIAM Journal on Scientific Computing*, 19(4):1188–1209, 1998.

[111] I. Maros and C. Mészáros. A repository of convex quadratic programming problems. *Optimization Methods and Software*, 11:671–681, 1999.

[112] The Mathworks. Matlab optimization toolbox user's guide, 2010.

[113] D. Meidner and B. Vexler. Adaptive space-time finite element methods for parabolic optimization problems. *SIAM Journal on Control and Optimization*, 46(1):116–142, 2007.

[114] C. Mészáros. The BPMPD interior point solver for convex quadratic problems. *Optimization Methods and Software*, 11(1):431–449, 1999.

[115] B. Morini. Convergence behaviour of inexact Newton methods. *Mathematics of Computation*, 68(228):1605–1613, 1999.

[116] MOSEK. The MOSEK optimization tools manual, version 6.0 (revision 85), 2010. URL http://www.mosek.com/.

[117] M.F. Murphy, G.H. Golub, and A.J. Wathen. A note on preconditioning for indefinite linear systems. *SIAM Journal on Scientific Computing*, 21(6): 1969–1972, 2000.

[118] K.G. Murty. Some NP-complete problems in quadratic and nonlinear programming. *Mathematical Programming*, 39:117–129, 1987.

[119] R.A. Nicolaides. Deflation of conjugate gradients with applications to boundary value problems. *SIAM Journal on Numerical Analysis*, 24(2): 355–365, 1987.

[120] S. Nilchan and C. Pantelides. On the optimisation of periodic adsorption processes. *Adsorption*, 4:113–147, 1998.

[121] J. Nocedal and S.J. Wright. *Numerical Optimization*. Springer Verlag, Berlin Heidelberg New York, 2nd edition, 2006.

[122] J.M. Ortega and W.C. Rheinboldt. *Iterative solution of nonlinear equations in several variables*. Academic Press, New York, 1970.

[123] C.C. Paige and M.A. Saunders. Solutions of sparse indefinite systems of linear equations. *SIAM Journal on Numerical Analysis*, 12(4):617–629, 1975.

[124] B.N. Parlett and W.G. Poole, Jr. A geometric theory for the QR, LU and power iterations. *SIAM Journal on Numerical Analysis*, 10:389–412, 1973.

[125] K.J. Plitt. Ein superlinear konvergentes Mehrzielverfahren zur direkten Berechnung beschränkter optimaler Steuerungen. Diploma thesis, Rheinische Friedrich–Wilhelms–Universität Bonn, 1981.

[126] L.S. Pontryagin, V.G. Boltyanski, R.V. Gamkrelidze, and E.F. Miscenko. *The Mathematical Theory of Optimal Processes*. Wiley, Chichester, 1962.

[127] A. Potschka. Handling path constraints in a direct multiple shooting method for optimal control problems. Diploma thesis, Universität Heidelberg, 2006. URL http://apotschka.googlepages.com/APotschka2006.pdf.

[128] A. Potschka, H.G. Bock, and J.P. Schlöder. A minima tracking variant of semi-infinite programming for the treatment of path constraints within direct solution of optimal control problems. *Optimization Methods and Software*, 24(2):237–252, 2009.

[129] A. Potschka, C. Kirches, H.G. Bock, and J.P. Schlöder. Reliable solution of convex quadratic programs with parametric active set methods. Technical Report 2010–11–2828, Heidelberg University, Interdisciplinary Center for Scientific Computing, Heidelberg University, Im Neuenheimer Feld 368, 69120 Heidelberg, Germany, 2010. URL http://www.optimization-online.org/DB_HTML/2009/02/2224.html.

[130] A. Potschka, A. Küpper, J.P. Schlöder, H.G. Bock, and S. Engell. Optimal control of periodic adsorption processes: The Newton-Picard inexact SQP method. In *Recent Advances in Optimization and its Applications in Engineering*, pages 361–378. Springer Verlag, 2010.

[131] A. Potschka, M.S. Mommer, J.P. Schlöder, and H.G. Bock. A Newton-Picard approach for efficient numerical solution of time-periodic parabolic PDE constrained optimization problems. Technical Report 2010–03–2570, Interdisciplinary Center for Scientific Computing (IWR), Heidelberg University, 2010. URL http://www.optimization-online.org/DB_HTML/2010/03/2570.html.

[132] S.M. Robinson. Perturbed Kuhn-Tucker points and rates of convergence for a class of nonlinear programming algorithms. *Mathematical Programming*, 7:1–16, 1974.

[133] Y. Saad. *Numerical methods for large eigenvalue problems*. Algorithms and Architectures for Advanced Scientific Computing. Manchester University Press, Manchester, 1992.

[134] Y. Saad. A flexible inner-outer preconditioned GMRES algorithm. *SIAM Journal on Scientific Computing*, 14(2):461–469, 1993.

[135] Y. Saad. *Iterative Methods for Sparse Linear Systems*. SIAM, Philadelpha, PA, 2nd edition, 2003.

[136] Y. Saad and M.H. Schultz. GMRES: A generalized minimal residual algorithm for solving nonsymmetric linear systems. *SIAM Journal on Scientific and Statistical Computing*, 7:856–869, 1986.

[137] S. Sager. Lange Schritte im Dualen Simplex-Algorithmus. Diploma thesis, Universität Heidelberg, 2001. URL http://mathopt.de/PUBLICATIONS/Sager2001.pdf.

[138] S. Sager. *Numerical methods for mixed–integer optimal control problems*. Der andere Verlag, Tönning, Lübeck, Marburg, 2005. URL http://mathopt.de/PUBLICATIONS/Sager2005.pdf.

[139] S. Sager, C. Barth, H. Diedam, M. Engelhart, and J. Funke. Optimization as an analysis tool for human complex problem solving. *SIAM Journal on Optimization*, 2011. Accepted.

[140] A. Schäfer, H. Mara, J. Freudenreich, C. Bathow, B. Breuckmann, and H.G. Bock. Large scale Angkor style reliefs: High definition 3D acquisition and improved visualization using local feature estimation. In *Proc. of 39th Annual Conference of Computer Applications and Quantitative Methods in Archaeology (CAA), Beijing, China*, 2011. Accepted.

[141] A.A.S. Schäfer. *Efficient reduced Newton-type methods for solution of large-scale structured optimization problems with application to biological and chemical processes*. PhD thesis, Universität Heidelberg, 2005. URL http://archiv.ub.uni-heidelberg.de/volltextserver/volltexte/2005/5264/.

[142] H. Schmidt-Traub, editor. *Preparative Chromatography of Fine Chemicals and Pharmaceuticals*. Harwell Report. Wiley-VCH, 2005.

[143] J. Schöberl and W. Zulehner. Symmetric indefinite preconditioners for saddle point problems with applications to PDE-constrained optimization problems. *SIAM Journal on Matrix Analysis and Applications*, 29(3):752–773, 2007.

[144] H. Schramm, S. Grüner, and A. Kienle. Optimal operation of Simulated Moving Bed chromatographic processes by means of simple feedback control. *Journal of Chromatography A*, 1006:3–13, 2003.

[145] H. Schramm, M. Kaspereit, A. Kienle, and A. Seidel-Morgenstern. Simulated moving bed process with cyclic modulation of the feed concentration. *Journal of Chromatography A*, 1006:77–86, 2003.

[146] L.F. Shampine and M.W. Reichelt. The MATLAB ODE suite. *SIAM Journal on Scientific Computing*, 18(1):1–22, 1997.

[147] G.M. Shroff and H.B. Keller. Stabilization of unstable procedures: the recursive projection method. *SIAM Journal on Numerical Analysis*, 30(4): 1099–1120, 1993.

[148] W. Squire and G. Trapp. Using complex variables to estimate derivatives of real functions. *SIAM Review*, 40:110–112, 1998.

[149] G.W. Stewart. Simultaneous iteration for computing invariant subspaces of non-Hermitian matrices. *Numerische Mathematik*, 25:123–136, 1976.

[150] Vidar Thomée. *Galerkin finite element methods for parabolic problems*, volume 25 of *Springer Series in Computational Mathematics*. Springer-Verlag, Berlin, 2nd edition, 2006.

[151] A. Toumi, S. Engell, M. Diehl, H.G. Bock, and J.P. Schlöder. Efficient optimization of Simulated Moving Bed Processes. *Chemical Engineering and Processing*, 46 (11):1067–1084, 2007.

[152] F. Tröltzsch. *Optimale Steuerung partieller Differentialgleichungen: Theorie, Verfahren und Anwendungen*. Vieweg+Teubner Verlag, Wiesbaden, 2nd edition, 2009.

[153] B.A. Turlach. QuadProg (quadratic programming routines), release 1.4, 1998. URL http://school.maths.uwa.edu.au/~berwin/software/quadprog.html.

[154] R. Tyson, S. R. Lubkin, and J. D. Murray. Model and analysis of chemotactic bacterial patterns in a liquid medium. *Journal of Biology*, 38:359–375, 1999.

[155] R. Tyson, S. R. Lubkin, and J. D. Murray. A minimal mechanism for bacterial pattern formation. *Proceedings of the Royal Society B: Biological Sciences*, 266:299–304, 1999.

[156] T.L. van Noorden, S.M. Verduyn Lunel, and A. Bliek. Optimization of cyclically operated reactors and separators. *Chemical Engineering Science*, 58:4114–4127, 2003.

[157] R.J. Vanderbei. LOQO: An interior point code for quadratic programming. *Optimization Methods and Software*, 11(1–4):451–484, 1999.

[158] A. Walther. A first-order convergence analysis of Trust-Region methods with inexact Jacobians. *SIAM Journal on Optimization*, 19(1):307–325, 2008.

[159] A. Walther, A. Kowarz, and A. Griewank. ADOL-C: A package for the automatic differentiation of algorithms written in C/C++. Technical report, Institute of Scientific Computing, Technical University Dresden, 2005.

[160] X. Wang. Resolution of ties in parametric quadratic programming. Master's thesis, University of Waterloo, Ontario, Canada, 2004.

[161] T. Warburton and M. Embree. The role of the penalty in the local discontinuous galerkin method for Maxwell's eigenvalue problem. *Computer Methods Applied Mechanical Engineering*, 195:3205–3223, 2006.

[162] A.J. Wathen. Realistic eigenvalue bounds for the Galerkin mass matrix. *IMA Journal of Numerical Analysis*, 7(4):449–457, 1987.

[163] R.C. Whaley, A. Petitet, and J.J. Dongarra. Automated empirical optimization of software and the ATLAS project. *Parallel Computing*, 27(1–2):3–35, 2001.

[164] E.P. Wigner. The unreasonable effectiveness of mathematics in the natural sciences. *Communications on Pure and Applied Mathematics*, 13:1–14, 1960. Richard Courant lecture in mathematical sciences delivered at New York University, May 11, 1959.

[165] J.H. Wilkinson. *The Algebraic Eigenvalue Problem*. Clarendon Press, Oxford, 1965.

[166] L. Wirsching. An SQP algorithm with inexact derivatives for a Direct Multiple Shooting method for optimal control problems. Diploma thesis, Universität Heidelberg, 2006.

[167] J. Wloka. *Partielle Differentialgleichungen: Sobolevräume u. Randwertaufgaben*. B.G. Teubner, Stuttgart, 1982.

[168] X. Zhang and Y. Ye. User's guide of COPL_QP, computational optimization program library: Convex quadratic programming, 1998. URL http://www.stanford.edu/~yyye/Col.html.

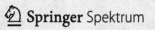